T0295470

Handbook on Renewable Energy and Green Technology

Handbook on
Renewable Energy and Green Technology

S.Pugalendhi
J.Gitanjali
R.Shalini
P. Subramanian

CRC Press is an imprint of the
Taylor & Francis Group, an **informa** business

First published 2024
by CRC Press
4 Park Square, Milton Park, Abingdon, Oxon, OX14 4RN

and by CRC Press
2385 NW Executive Center Drive, Suite 320, Boca Raton FL 33431

CRC Press is an imprint of Informa UK Limited

British Library Cataloguing-in-Publication Data
A catalogue record for this book is available from the British Library

ISBN13: 9781032711898 (hbk)
ISBN13: 9781032711904 (pbk)
ISBN13: 9781032711911 (ebk)

DOI: 10.4324/9781032711911

Typeset in Adobe Caslon Pro
by Elite Publishing House, Delhi

Contents

Preface

As the world population grows and places more demand on limited fossil fuels, renewable energy becomes more relevant as part of the solution to the impending energy dilemma. Renewable energy supplies are of ever increasing environmental and economic importance in all the countries. A wide range of renewable energy technologies have been established commercially and recognized as growth industries by the governments. Energy supply from renewable is an essential component of every nation's strategy, especially when there is responsibility for the environment and for sustainability.

"Handbook on Renewable Energy and Green Technology" is a collection and compilation of various principles of renewable energy technologies and explores how we can use the sun, wind, biomass, geothermal, tidal and water resources to generate energy in more sustainable form. Each chapter begins with fundamental theory behind each technology illustrated with clear figures to understand the principle and applications in an easy way. It also explains the fundamentals of energy, including the transfer of energy, as well as the limitations of natural resources. Starting with solar and wind energy, the text illustrates how energy from the sun, wind and water are transferred and converted into electricity.

Other chapters cover methods of energy conversion, biomass energy, biofuel production and other new and renewable sources of energy such as geothermal, hydro, tidal, and ocean energy.

This book will be a basic guide for the beginners those who are new in the field of renewable energy especially, under graduate and post graduate degree students. Moreover, since many practicing scientists and engineers will not have had a general training in renewable energy, this book will have wider use beyond colleges and universities. The whole book is structured to share common material and to relate aspects together. Therefore, the book is intended both for basic study and for application.

Authors

About the Authors

S.Pugalendhi

He is currently serving as the Emeritus Professor in the Department of Renewable Energy Engineering, Agricultural Engineering College and Research Institute, Tamil Nadu Agricultural University, Coimbatore. He has teaching experience of about 30 years and published more than 28 articles in National and International Journals. He has edited 9 books so far in the field of Agriculture Engineering both in English and Tamil languages. He is specialized in the area of converting organic wastes into biogas and developed different types of biogas plants, biogas cook stoves and biogas lamp at low costs. From 1994 to till now, he has conducted more than thousands of biogas training programs covering most of the villages in Tamil Nadu and constructed nearly 1000 Deenabandhu biogas plants in the successful way through government schemes.

J.Gitanjali

She is currently working as a Teaching Assistant in the Department of Farm machinery and Power Engineering, Agricultural Engineering College and Research Institute, Tamil Nadu Agricultural University, Coimbatore. She was awarded for Best M.Tech Student and Best Ph.D Student during the year 2012 and 2016 respectively from TNAU. She has published 12 research papers in National and International Journals. She is specialized in the area of thermochemical conversion technologies especially combustion and plasma gasification.

R.Shalini

She is currently working as a Scientist cum Research Co-oridnator (India), Capture Bank AS. During 2013, she got National Award for Best M.Tech Thesis in the field of Agricultural Engineering by Indian Society for Technical Education (ISTE), New Delhi. Doctoral thesis investigated on designing a low cost, lab-scale hydrothermal carbonization reactor (1000 ml) for converting wet/dry agricultural renewable biomass residues into novel carbon-rich solid product called hydrochars. She has published 4 research paper in reputed journal, 4 popular articles and edited 2 books in Tamil. She is specialized in field of thermochemical conversion of biomass, especially biochar production through pyrolysis process.

P. Subramanian

He is currently working as the Professor and Head in the Department of Agricultural Engineering, Agricultural College and Research Institute, Tamil Nadu Agricultural University, Madurai. He has teaching experience of about 25 years. He has published 10 research papers in International Journals and 16 research papers in National Journals in addition to 2 International and 25 National seminar papers. He is specialized in the area of biomass gasification, biofuel production technologies and energy auditing and management.

Chapter - 1

Importance of Renewable Energy Sources

1.1. Importance of Energy

Energy is the driving force in the development of economic growth, progress and prosperity of any nation. Energy demand in agriculture, industrial, commercial and household sectors has increased tremendously and placed enormous pressure on its resources. The economic growth level of any nation will be indicated by the availability and consumption levels of energy. India's substantial economic growth is placing enormous demand on its energy resources. The depleting resources and increasing pollution of environment due to energy has necessities optimum use of its resources; which in turn requires proper energy planning to achieve energy security.

The fossil fuel supply viz. Coal, Petroleum, Natural gas will be depleted in a few hundred years. Industrialization, urbanization, population growth, economic growth, improvement in per capita consumption of electricity, depletion of coal reserve, increasing import of coal, crude oil and other energy sources and the rising concern over climate change have put India in a critical position. It has to take a tough stance to balance between economic development and environmental sustainability.

Therefore, alternate or renewable sources of energy have to be developed to meet future energy requirement. To reduce dependency on imports, India's strategy is the encouragement of the development of various forms of renewable sources of energy by the use of incentives by the Federal and State governments including the use of nuclear energy, promoting wind farms, solar and biomass energy.

The following important key points have to be considered to promote energy security:

- Easy availability of energy helps in the process of industrialization in a country
- Production of energy leads to the efficient utilization of natural resources. For example, solar energy, wind energy and hydro-electricity power can be generated by using sunlight, wind and water resources respectively
- Scope of employment opportunities can be possible with the process of industrialization that is possible with easy availability of energy/power sources
- Easy availability of energy is required for the expansion of infrastructural development in a country
- Income of a country can be raised with the expansion of the power sector. It also helps to achieve economic self sufficiency

1.2. World energy scenario

Renewables are the world's fastest-growing energy source over the projection period between 2012 and 2040. Annual renewable energy consumption increases by an average of 2.6%. Nuclear power is the world's second fastest-growing energy source, with an annual consumption increasing by 2.3 % over that period. Even though consumption of non fossil fuels is expected to grow faster than consumption of fossil fuels, fossil fuels still account for 78% of energy use in 2040. Annual global natural gas consumption increases by 1.9% Abundant natural gas resources and robust production including rising supplies of tight gas, shale gas and coal bed methane contribute to the strong competitive position of natural gas. Although, liquid fuels mostly petroleum based remain the largest source of world energy consumption, the liquids share of world marketed energy consumption falls from 33% in 2012 to 30% in 2040. Contributing to the decline are rising oil prices in the long term, which lead many energy users to adopt more energy-efficient technologies and to switch away from liquid fuels when feasible.

World use of petroleum and other liquid fuels grows from 90 million barrels per day in 2012 to 100 million barrels per day in 2020 and to 121 million barrels per day in 2040. Worldwide natural gas consumption is projected to increase from 120 trillion cubic feet in 2012 to 203 trillion cubic feet in 2040. Coal is the world's slowest-growing energy source rising by an average 0.6% per year, from 153 quadrillion Btu in 2012 to 180 quadrillion Btu in 2040. World net electricity generation increases by 69% from 21.6 trillion kWh in 2012 to 25.8 trillion kWh in 2020 and to 36.5 trillion kWh in 2040 (International Energy Outlook, 2016).

Renewables are the fastest growing source of energy for electricity generation, with annual average increase of 2.6% from 2012 to 2040. In contrast to renewables, coal-fired net generation increases by 0.8%. Electricity generation from nuclear power worldwide increases from 2.3 trillion kWh in 2012 to 3.1 trillion kWh in 2020 and to 4.5 trillion kWh in 2040 Total world consumption of marketed energy expands from 549 quadrillion British thermal units (Btu) in 2012 to 629 quadrillion Btu in 2020 and to 815 quadrillion Btu in 2040 resulting in 48% increase from 2012 to 2040.

1.3. Indian Energy Scenario

1.3.1. Renewable energy potential

India is blessed with vast renewable energy sources such as solar, wind, biomass and hydro energy. In India, total potential for renewable power generation is estimated as 1096 GW (March 2018). This includes solar power potential of 748990 MW (68.33%), wind power potential of 302251 MW (27.58%) at 100m hub height, SHP (small-hydro power) potential of 19749 MW (1.80%), Biomass power of 17,536 MW (1.60%), 5000 MW (0.46%) from bagasse-based cogeneration in sugar mills and 2554 MW (0.23%) from waste to energy (MoSPI, 2019).

1.3.2. Installed capacity of energy sources

In India, electric power generation is the largest source of greenhouse gas emissions and accounts for 48 per cent of the carbon emitted. These concerns, point towards more rational energy use strategies. Perhaps, renewable energy based technologies, functioning in a sustainable manner is the way forward. The country having diverse climatic region is well endowed with renewable energy resources. The current installed power generation capacity in India is about 403.76 GW which comprises of 58.50 % from thermal, 1.70 % from nuclear, 11.60 % from hydro and 28.30% from renewable sector. Fig.1.1. shows the total installed capacity of renewable and non-renewable sources of energy (MoP, 2022).

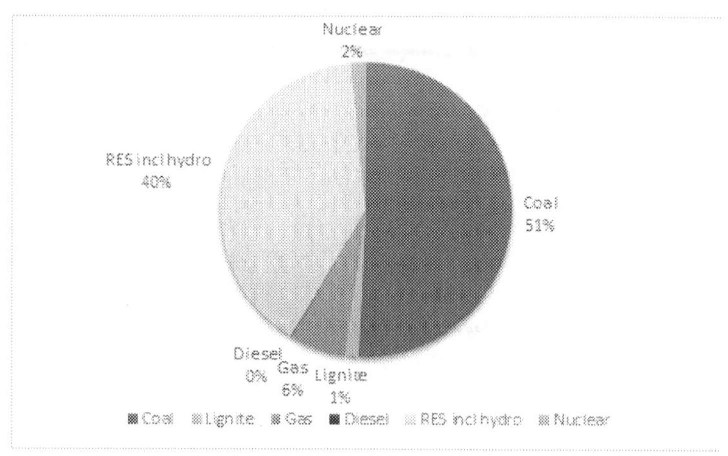

Fig. 1.1. Indian Energy Scenario

1.3.3. Tamil Nadu energy scenario

Tamil Nadu has a total installed capacity of renewable and non-renewable energy of 23,104.15 MW. Of which, renewable energy sources contribute about 8432.15 MW. Tamil Nadu has 40% of the entire wind energy potential of India. Tamil Nadu Solar Energy Policy aims over 3000 MW by 2015. Due to the astronomical increase in energy demand in recent years, the state has a power deficit of about 14.9%. To meet the ever-increasing energy demand, TNEB has proposed a number of next-generation projects to be constructed over the next 5 years (TEDA, 2018).

1.4. Classification of Energy Sources

Energy can be classified into several types based on the following criteria:

a. Primary and Secondary energy sources

b. Commercial and Non-commercial energy sources

c. Renewable and Non-renewable energy sources

1.4.1. Primary and Secondary energy sources

Primary energy sources are those that are either found or stored in nature. Common primary energy sources are coal, oil, natural gas, and biomass (such as wood). Other primary energy sources available include nuclear energy from radioactive substances, thermal energy stored in earth's interior, and potential energy due to earth's gravity. Primary energy sources are mostly converted in industrial utilities into secondary energy sources; for example coal, oil or gas converted into steam and electricity. Primary energy can also be used directly. Some energy sources have non-energy uses, for example coal or natural gas can be used as a feedstock in fertilizer plants.

1.4.2. Commercial and Non commercial energy sources

The energy sources that are available in the market for a definite price are known as commercial energy. By far the most important forms of commercial energy are electricity, coal and refined petroleum products. Commercial energy forms the basis of industrial, agricultural, transport and commercial development in the modern world. In the industrialized countries, commercialized fuels are predominant source not only for economic production, but also for many household tasks of general population. Examples: Electricity, lignite, coal, oil, natural gas etc.

The energy sources that are not available in the commercial market for a price are classified as non-commercial energy. Non-commercial energy sources include fuels such as firewood, cattle dung and agricultural wastes, which are traditionally gathered, and not bought at a price used especially in rural households. These are also called traditional fuels.

Non-commercial energy is often ignored in energy accounting. Examples: Firewood, agro waste in rural areas; solar energy for water heating, electricity generation and drying grain, fish and fruits, animal power for transport, threshing, lifting water for irrigation and crushing sugarcane and wind energy for lifting water and electricity generation.

1.4.3. Renewable and Non-Renewable energy sources

Renewable energy is the energy obtained from sources that are essentially inexhaustible. Examples of renewable resources include wind, solar, tidal and hydroelectric. The most important feature of renewable energy is that it can be harnessed without the release of harmful pollutants. Non-renewable energy is the conventional fossil fuels such as coal, oil and gas, which are likely to deplete with time.

1.5. Renewable Energy Achievements in India

The current installed capacity of renewable energy gadgets for power production given by the Ministry of New and Renewable Energy for the year 2016 is listed in Table 1.1.

1.6. Solar Energy

India is endowed with vast solar energy. The solar radiation of about 1,74,000 TW is incident over its land mass with average daily solar power potential of 0.25 kWh per m^2 and overall potential of 19 GW. The installed capacity is 4579.24 MW in which 3743 MW is from solar photo voltaic and 836.24 MW is from solar thermal sector. India expects to install an additional 10,000 MW by 2017 and a total of 1,00,000 MW by 2022 (Energy Statistics, 2015).

1.6.1. Solar energy schemes in India

The government of India is promoting the use of solar energy through various strategies. The Jawaharlal Nehru National Solar Mission (JNNSM) was launched on the 11th January, 2010 by the honourable Prime Minister of India. The objective of this mission under the brand 'Solar India' is to establish India as a global leader in solar energy, by creating the policy conditions for its diffusion across the country as quickly as possible. The target of this mission is to achieve 20,000 MW by 2022 with a budget of 1380 crores. MNRE provides 70% subsidy on the installation cost of a solar photovoltaic power plant in North-East states and 30% subsidy on other regions. MNRE along with NABARD implemented a scheme under JNNSM to promote solar home lighting systems to rural areas. These scheme offers 30-40% subsidies to promote solar PV home lighting systems (MNRE, 2016). The solar energy can be effectively harnessed using two basic technologies such as solar thermal and photovoltaic conversion systems which have been explained in a separate chapter.

Table1.1. Renewable energy achievement in India (Source: MNRE, 2022)

Sector	Cumulative Achievements (as on 31.07.2022)
Wind Power	40893.33
Solar Power*	57973.80
Small Hydro Power	4887.90
Biomass (Bagasse) Cogeneration	9433.56
Biomass(non-bagasse) Cogeneration	772.05
Waste to Power	223.14
Waste to Energy (off-grid)	253.61
Total	**114437.39**

1.7. Wind Energy

Winds are caused by the uneven heating of the atmosphere by the sun, the irregularities of the earth's surface and rotation of the earth. The terms wind energy or wind power describes the process by which the wind is used to generate mechanical power or electricity. Wind turbines convert the kinetic energy in the wind into mechanical power. This mechanical power can be used for specific tasks (such as grinding grain or pumping water) or a generator can convert this mechanical power into electricity.

1.8. Biomass Energy

Biomass is defined as renewable energy sources derived from organic carbonaceous materials originating from plants, animals and micro-organisms that contain energy in a chemical form that can be converted into fuel. Biomass energy is produced by green plants by photosynthesis in the presence of sun light. Biomass is the fourth largest source of energy and most important fuel worldwide after coal, oil and natural gas.

$$6\,CO_2 + 6\,H_2O + \text{Light energy} \text{-------}\rightarrow 6\,O_2 + C_6H_{12}O_6$$

Among renewable energy options, agricultural and biomass residues can be used as raw materials for the production of solid, liquid and gaseous products. Biomass has been identified as the possible sustainable renewable source, which can fulfill the energy demand without polluting the environment. Use of renewable and carbon capture and storage (CCS) technologies could reduce global level coal consumption and also global CO_2 emissions from coal could be reduced to 5.6 Gt by 2035.

Biomass has an estimated annual potential of about 500 million tonnes which meets 33% of the primary energy needs. It is considered as one of the important renewable sources for mankind as fuel wood which meets about 75% of the rural energy needs and rural population constitutes 70% of the total population (MNRE, 2016). Conversion of biomass into energy is undertaken using two main process technologies namely thermochemical and biochemical. Thermo chemical conversion process includes combustion, pyrolysis and gasification whereas biochemical conversion process includes anaerobic digestion and fermentation.

1.9. Other New and Renewable Energy Sources

The other new and renewable sources of energy include hydro, geothermal, tidal, wave and OTEC.

1.10. Advantages of Renewable Sources of Energy

- Inexhaustible, replenishable and abundant in nature
- Reliable
- Stabilized energy price
- Eco-friendly
- Free of pollution and costs
- Mitigates green house emissions such as carbon dioxide (CO_2), methane (CH_4) and nitrous oxide (N_2O)

1.11. Limitations

- Higher capital costs
- Energy is unpredictable and inconsistent since depends on the season

1.12. Energy Management in Agriculture

The energy sources used in agriculture are mainly of commercial energy, which is nonrenewable in origin, and its consumption is about 80 per cent. Increased usage of these sources will not only deplete the reserve, but also increases the production cost. The prime concern is maximizing productivity per unit area with minimum of energy input and cost. Energy utilized in agriculture varies widely according to the cropping pattern, type of farm activities and the level of mechanization.

The study on energy consumption pattern in crop production helps in identifying major energy consuming operations or sources. This also helps in studying the possibilities of conservation of energy and also optimal allocation of available energy sources in crop production system.

The energy consumption in terms of operation and source wise can be calculated. The operations considered are tillage, sowing, bund making, weeding, fertilizer application, spraying, harvesting and threshing. The total energy consumed in each operation was calculated by summing all the individual energy sources used for each operation. Likewise, each source energy consumption (human, animal, diesel, electricity, machinery, seed, fertilizer, FYM and chemicals) for all the operations are also calculated.

Further energy management will suggest the time and amount of application of various inputs in view of considerable increase in yield. Hence, through energy management studies in agriculture, the available energy sources can be utilized effectively and efficiently for higher productivity.

The quantity of energy requirement for crop production varies between 5000 and 50000 MJ ha^{-1} depends on crops and cultivation practices. The disadvantages of conventional energy sources are high production cost, heavy transmission and distribution loss, cause of global warming, greenhouse effect and acid rain, uncertainty in availability, biggest threat for environment and public health, very expansive to be maintained, stored and transmitted and exhaustible in nature.

Chapter - 2

Methods of Energy Conversion Technology

A wide variety of conversion technologies is available for manufacturing premium fuels from biomass. Conversion of biomass into biofuel can be achieved by two methods, namely thermochemical and biochemical conversion technologies which are briefly illustrated in Fig.2.1.

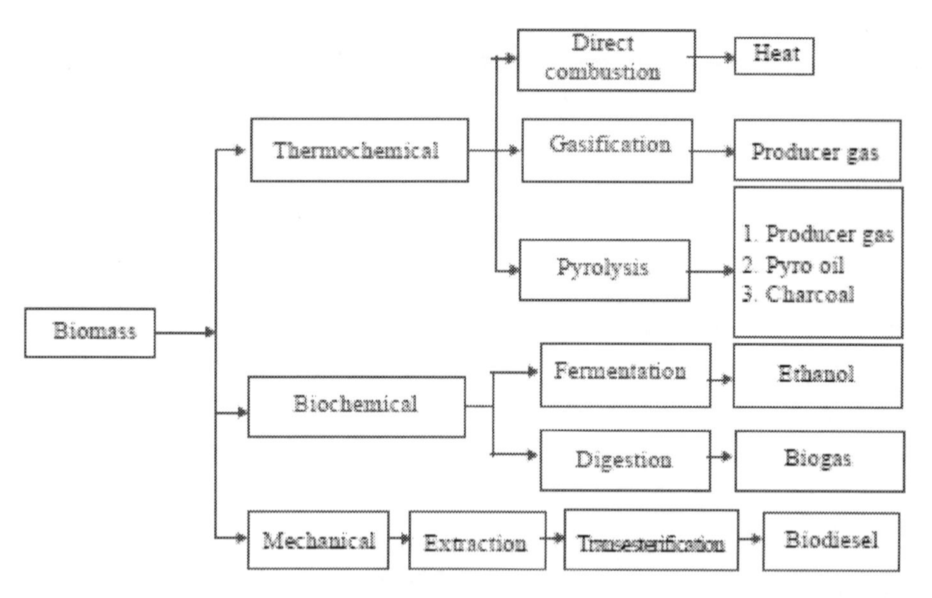

Fig. 2.1. Energy conversion routes

Each biomass resources such as wood, dung, vegetable wastes, agro residues can be treated in many different ways to provide a wide spectrum of useful products. Biomass can be dried and burnt to provide heat or converted into low calorific value gas by pyrolysis process. Alternatively, it can be stirred into slurry and digested to yield methane. Likewise, liquid and gaseous fuels such as methanol and methane can be manufactured by several different routes and from a variety of biomass.

The choice of process is determined by a number of factors such as location and type of the feed stocks, physical properties, economics and availability of a suitable market for the product.

2.1. Thermochemical Method of Conversion

Thermal conversion processes use heat to convert biomass into solid, liquid and gaseous products depending on the technology adopted. The basic thermochemical technologies are combustion, pyrolysis and gasification which produce different end products.

2.1.1. Combustion

Combustion refers to the rapid oxidation of fuel accompanied by the production of heat, or heat and light. Complete combustion of a fuel is possible only in the presence of an adequate supply of oxygen. Oxygen (O_2) is one of the most common elements on earth making up 20.9%(v) of atmospheric air. Rapid fuel oxidation results in large amounts of heat. Solid or liquid fuels must be changed to a gas before they will burn. Usually heat is required to change liquids or solids into gases. During combustion process, if carbon monoxide (CO) is released, it is called as incomplete combustion and if carbon dioxide (CO_2) is released, it is called as complete combustion.

2.1.2. Pyrolysis

Pyrolysis is defined as the thermal degradation of biomass at temperature between 250 – 900 °C in the absence of oxygen to yield solid product called charcoal or biochar, liquid product called biooil and gaseous product called syngas. The pyrolysis process is classified into slow, intermediate and fast pyrolysis depending on its reaction temperature, residence time and type of product required.

» Charcoal or biochar can be used as a substitute for fuel and as a soil conditioner

» Bio-oil can be used as an alternate liquid fuel for engine and other heating applications

» Syngas can be used for thermal application and power generation

2.1.3. Gasification

Gasification is a thermo-chemical process in which, biomass is thermally degraded with restricted amount of air / oxygen up to 900 °C to yield a mixture of combustible gases consisting carbon monoxide, hydrogen and traces of methane called producer gas. The reactor where the gasification takes place is called as gasifier. The producer gas can be used for both thermal application and power generation.

2.2. Biochemical Method of Conversion

Biochemical conversion of organic waste materials involves the activity of microbes, to get converted into useful products. The biochemical technologies are as follows:

2.2.1. Aerobic method of conversion

The aerobic method of conversion of biomass is composting. Composting is a bio-process, in which diverse and mixed group of micro-organisms breakdown organic materials and transform in to humus like substances in aerobic conditions. The organic material will normally have an indigenous mixed population of microorganisms derived from the atmosphere, water or soil. When the moisture content of the wastes is brought to a suitable level and the mass aerated, microbial action speeds up. As well as oxygen and moisture, the microorganisms require for their growth and reproduction a source of carbon (the organic waste), macronutrients such as nitrogen, phosphorus and potassium and certain trace elements. In attacking the organic matter, the microorganisms reproduce themselves and liberate carbon dioxide, water, other organic products and energy. Some of the energy is used in metabolism; the remainder is given off as heat. The end product, compost is made up of the more resistant molecules of the organic matter, breakdown products, dead and some living microorganisms, together with products of biochemical reaction.

2.2.2. Anaerobic method of conversion

The biomass is converted into energy in the form of biogas and liquid biofuels through anaerobic methods. They are as follows:

2.2.2.1. Biogas

Anaerobic digestion involves the microbial digestion of biomass. The process takes place at low temperature upto 65 °C and requires a moisture content of at least 80%. It generates a gas called biogas or gobar gas which consisting mostly of 55-65% methane (CH_4) and 35 – 45% carbon dioxide (CO_2) with minimum impurities of hydrogen sulfide. The biogas can be burned directly or upgraded to synthetic natural gas by removing the CO_2 and the impurities. The residue may consist of protein-rich sludge that can be used as animal feed and liquid effluents that are biologically treated by standard techniques and returned to the soil.

2.2.2.2. Liquid biofuels

Depleted supplies of fossil fuel, regular price hikes of gasoline and environmental damage have necessitated the search for economic and eco-friendly alternative of gasoline. By 2020, India's dependency on oil imports is projected to go beyond 90% from 80% at current level. So it is time to shift for alternate fuels like biodiesel and bioethanol which are produced from biomass, renewable and abundantly available source.

Ethanol is produced through fermentation of sugar, starch and lignocellulosic materials. Fermentation is the breakdown of complex molecules in organic compound under the influence of ferment such as yeast, bacteria, enzymes, etc. fermentation is well established and widely used technology for the conversion of grains and sugar crops into ethanol. Ethanol can be mixed with gasoline to produce gasohol (90% gasoline and 10% ethanol).

Butanol was tradionally produced by ABE fermentation - the anaerobic conversion of carbohydrates by strains of *Clostridium* into acetone, butanol and ethanol. Acetone–butanol–ethanol (ABE) fermentation is a process that uses bacterial fermentation to produce acetone, n-Butanol, and ethanol from carbohydrates such as starch and glucose.

2.3. Chemical Method of Conversion

Biodiesel is a chemical method of conversion of oil into diesel. Biodiesel is a variety of ester-based oxygenated fuels derived from natural, renewable biological sources such as vegetable oils through transesterification process. Biodiesel can be used as an alternate fuel for diesel engines.

2.4. Characterization of Feedstock

The different properties such as physical properties, proximate analysis, ultimate analysis, thermal properties and biochemical analysis of the solid and liquid wastes are determined using standard methods.

2.4.1. Physical properties

a. Bulk density

Bulk density is the main physical property for biomass. It is determined by weighing the biomass sample filled in a vessel of known volume and calculating the ratio of the weight of the biomass sample to the volume of the vessel.

$$\text{Bulk density } (\rho), \text{kg/m}^3 = \frac{W}{V}$$

where,

W = Weight of the biomass sample, kg

V = Volume of the vessel, m^3

b. True density and porosity

True density is determined by calculating the difference between bulk density and porosity. The porosity is determined by the device is called air picno meter. It is expressed in percentage (%).

c. Moisture content

Using hot air oven (Fig.2.2.) moisture content of the sample is determined. By oven method (ASTM, E-871) about 5 g of powdered sample is taken in a petri dish and kept inside the oven at a temperature of 103 ± 5°C for 24 h. The moisture content is calculated by measuring weight of moisture evaporated from the sample (W_2) and the original sample weight (W_1). The formula is given below.

$$\text{Moisture content (\%)} = \frac{W_1 - W_2}{W_1} \times 100$$

where,

W_1 = Initial weight of the biomass sample, g

W_2 = Final weight of the biomass sample, g

Fig. 2.2. Hot air oven

d. pH

The pH of samples is estimated using the electrometric method (APHA, 1999). It is measured by pH meter having glass electrode and temperature probe as shown in Fig.2.3. A 100 mL of the sample is taken in a clean beaker. The electrode is dipped into the beaker containing waste water sample and the value is noted.

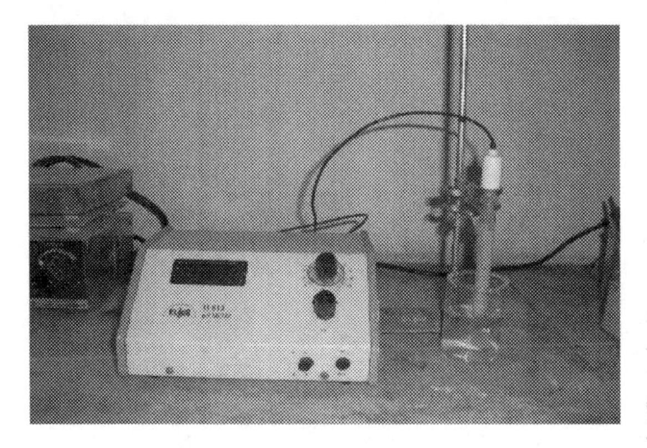

Fig.2.3. pH experiment setup

e. Total solids

The total solid (TS) is determined by the procedure outlined by APHA 1999, 2450 B, a standard method for examination of waste water.

A measured volume of well mixed sample is taken to a pre weighed dish (A) and evaporated to bone-dry condition in hot air oven. The evaporated sample is dried for one hour in the oven at 103 - 105 °C. The dish is then cooled in desiccator and weighed. The process of drying, cooling and weighing was repeated, until concordant value of mass is obtained.

$$\text{Total solids, (mg/L)} = \frac{(B - A) \times 1000}{\text{Sample volume, mL}}$$

Where,

A = Mass of dish, mg

B = Mass of dish + Dried residue, mg

f. Volatile solids

The residue obtained in TS determination was ignited in a muffle furnace at a temperature of 550 ± 50 °C for 30 minutes, in a closed porcelain crucible (APHA 1999, Standard Methods, Section 2540 C). After partial cooling, the crucible with residue is transferred to the desiccator for final cooling. The cycle of igniting, cooling and weighing is continued until concordant values were obtained.

$$\text{Volatile solids, (mg/L)} = \frac{(A - B)}{\text{Sample volume, mL}} \times 1000$$

Where,

A = Mass of residue + crucible before ignition, mg

B = Mass of residue + crucible after ignition, mg

g. Total suspended solids (TSS)

A well-mixed sample is filtered through a weighed standard glass-fibre filter and the residue retained on the filter is dried to constant mass in an oven at 103 - 105°C (APHA 1999, Standard Methods, Section 2540 D). The increase in mass over empty dish represents the mass of Total suspended solids.

$$\text{Total suspended solids, (mg/L)} = \frac{(D - C) \times 1000}{\text{Sample volume, mL}}$$

Where,

C = Mass of dish + filter, mg

D = Mass of dish + filter + dried residue, mg

h. Volatile suspended solids (VSS)

The residue obtained in TSS determination is ignited in a muffle furnace at a temperature of 550 ± 50°C for 30 minutes, in a closed porcelain crucible (APHA 1999, Standard Methods, Section 2540 E). After partial cooling, the crucible with the residue is transferred to the desiccator for final cooling. The cycle of igniting, cooling and weighing is continued until concordant values are obtained.

$$\text{Volatile suspended solids, (mg/L)} = \frac{(E - F) \times 1000}{\text{Sample volume, mL}}$$

Where,

E = Mass of dish + filter + dried residue before ignition, mg

F = Mass of dish + filter + dried residue after ignition, mg

2.4.2. Proximate analysis

Proximate composition such as volatile matter, ash content and fixed carbon content of the biomass species is calculated by using the procedure and formula given below. Mostly proximate analysis is determined on dry basis (%).

a. Volatile matter

The volatile matter is determined using muffle furnace (ASTM, E-872). To measure the volatile content, known content of dried sample is taken in a closed crucible and kept inside the muffle furnace (Fig.2.4) at 650 °C for six minutes and again at 750°C for another six minutes. The loss in weight of the sample is found out and the % of volatile matter is calculated as,

$$\text{Volatile matter (\%)} = \frac{\text{Loss in weight of the sample, g}}{\text{Weight of moisture free sample, g}} \times 100$$

Fig.2.4. Muffle furnace

b. Ash content

The ash content of the selected biomass is found out (ASTM, E-830) by taking a known quantity of dried biomass sample in an open crucible and keeping it in a muffle furnace at about 750°C up to reaching a standard weight. The ratio between the remaining weight of biomass in the crucible and the sample taken is the fraction of ash content of tested material.

$$\text{Ash content (\%)} = \frac{\text{Weight of ash formed, g}}{\text{Weight of dried sample, g}} \times 100$$

c. Fixed carbon

The fixed carbon of biomass sample was calculated by subtracting the sum of ash content (%) and volatile matter (%) from 100. The fixed carbon is the residue left after removing the volatile matter and the ash from the substance.

$$\text{Fixed Carbon, \%} = 100 - (\text{Volatile matter (\%)} + \text{Ash content (\%)})$$

2.4.3. Ultimate analysis

Elemental composition of biomass is an important property, which defines the energy content and determines the clean and efficient use of the biomass materials. However, the ultimate analysis requires very expensive equipment like elemental analyzer and highly trained analysts. The ultimate analysis of the biomass and other samples are determined the fractions by weight of the composition in carbon, hydrogen, nitrogen and oxygen is calculated using elemental analyzer (Fig.2.5.).

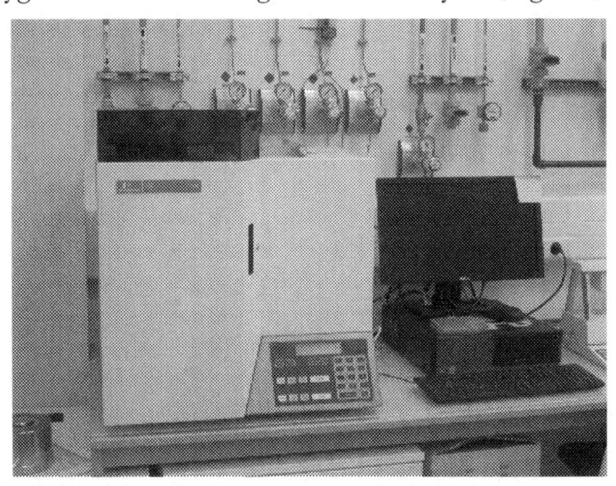

Fig.2.5. Elemental analyzer

Elemental C, H, N and O content is estimated using correlation analysis suggested by (Parikh *et al.*, 2007) from proximate analysis. The correlation formulae used to find elemental composition is as follows:

C - 0.637 FC + 0.455 VM

H - 0.052 FC + 0.062 VM

O - 0.304 FC + 0.476 VM

Where,

C - Carbon, %

H - Hydrogen, %

O - Oxygen, %

FC - Fixed Carbon, % and

VM - Volatile Matter, %

2.4.4. Thermal properties

a. Calorific value

Calorific value is the major thermal property in the study of biomass for any fuel applications. Heating value or calorific value is the heat released by the fuel under ideal combustion conditions. There are basically two types of calorific value namely, higher heating value (HHV) and lower heating value (LHV).

The Higher Heating Value (HHV) is the total amount of heat in a sample of fuel - including the energy in the water vapor that is created during the combustion process, when the combustion process takes place at constant volume and water formed during the process is condensed and the latent heat of vaporization of water is also accounted. The heating value determined by using Bomb calorimeter is the Higher Heating Value (HHV) or otherwise called as Gross Calorific Value (GCV).

In almost all the thermo-chemical conversion processes, such as gasification and pyrolysis, the process occurs at constant pressure and vapors leave the process as such without getting condensed. The heating value under these conditions is called as Lower Heating Value (LHV) or Net Calorific Value (NCV). The Lower Heating Value (LHV) is the amount of heat in a sample of fuel minus the energy in the combustion water vapor.

The Lower Heating Value is always less than the Higher Heating Value for a fuel. The suggested equation for calculating Lower Heating Value (LHV) from Higher Heating Value (HHV) is,

$$\text{Lower Heating Value} = \text{Higher Heating Value} - 2.26 \text{ MJ/kg}$$

Bomb calorimeter is used to determine calorific value of both solid and liquid samples. The calorific value of the gaseous products is determined using Junkers gas calorimeter. The unit for solid and liquid fuels is kJ/kg or MJ/kg and for gaseous fuels, it is kJ/Nm^3 or MJ/Nm^3.

Description of bomb calorimeter

The bomb calorimeter consists of a strong stainless steel bomb, inside which the fuel sample is burnt. The bomb is provided with an oxygen inlet valve and two stainless steel electrodes. To one of the electrodes, a small ring is attached. In this ring, a stainless steel crucible is placed. The bomb is placed in a copper calorimeter which contains three fourth of water. The copper calorimeter is provided with a Beckmann thermometer and a stirrer. The copper calorimeter is surrounded by an air jacket and water jacket.

Working principle of bomb calorimeter

A known weight of fuel sample is taken in the crucible. A fine magnesium wire, touching the fuel sample, is then stretched across the electrodes. The bomb lid is tightly screwed.

The bomb is filled with oxygen at 25 atm pressure. The bomb is kept in the copper calorimeter. Initial temperature of water in the calorimeter is noted after thorough stirring. The electrodes are then connected to 6V battery. The current is switched on and the fuel sample in the crucible burnt with the evolution of heat. The heat liberated due to the combustion of the sample raises the temperature of water in the calorimeter (Fig.2.6.).

Fig.2.6. Bomb calorimeter

The final temperature of water is noted. The calorific value of the fuel is calculated as follows.

$$CV_F, \text{cal/g} = \frac{(W + w)T}{M}$$

where,

CV_F - Calorific value, cal/g
W - Water equivalent of calorimeter assembly, cal/°C
T - Rise in temperature, °C
M - Mass of sample burnt, g

b. Thermo Gravimetric Analyzer

Thermo Gravimetric Analysis / Thermal Gravimetric Analysis (TGA) is a type of testing that is performed on biomass or other residue samples to determine changes in weight in relation to change in temperature. Such analysis relies on a high degree of precision in three measurements namely:

1. Weight,
2. Temperature and
3. Temperature change.

A derivative weight loss curve can be used to predict the point at which weight loss is more apparent. The thermo gravimetric analyzer (Fig.2.7.) consists of a high-precision balance with a pan (generally Platinum) loaded with the sample. The pan is placed in a small electrically heated oven with a thermocouple to accurately measure the temperature. The atmosphere may be purged with an inert gas to prevent oxidation or other undesired reactions.

Fig.2.7. Thermo Gravimetric Analyzer

Analysis is carried out by raising the temperature gradually and plotting weight in percentage against temperature. The testing temperature is 1000 °C or greater. After the data are obtained, curve smoothing and other operations may be done such as to find the exact points of inflection. A TGA graph of coconut shell biomass is given in Fig.2.8.

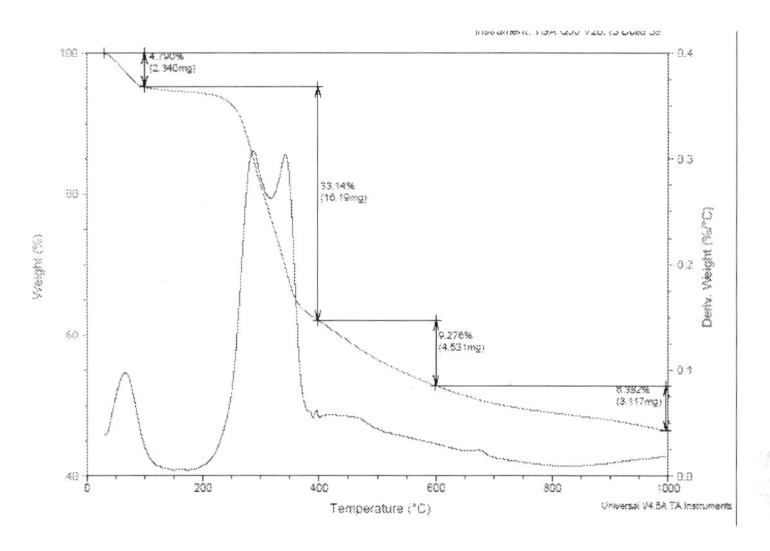

Fig. 2.8. TGA graph for coconut shell biomass

c. Energy density

Energy density is calculated using bulk density and calorific value of biomass feedstock.

$$\text{Energy Density, MJ/m}^3 = BD \times CV$$

Where,

BD = Bulk Density of the biomass, kg/m^3

CV = Calorific Value of the biomass, MJ/kg

2.4.5. Biochemical analysis

a. Biological Oxygen Demand (BOD)

The BOD test (Fig.2.9.) is carried out by diluting the sample with de-ionized water with added nutrients, saturated with oxygen, inoculating it with a fixed aliquot of seed, measuring the dissolved oxygen and sealing the sample (to prevent further oxygen dissolving in). The sample is kept at 20 °C in the dark to prevent photosynthesis (and thereby the addition of oxygen) for five days and the dissolved oxygen is measured again. The difference between the final DO and initial DO is the BOD.

- Undiluted: BOD = Initial DO - Final DO
- Diluted: BOD = (Initial DO - Final DO) - BOD of blank) x Dilution Factor

Fig.2.9. BOD experiment set up

b. Chemical oxygen demand (COD)

The COD of the samples are analyzed by Dichromate method (APHA, 1989). The diluted sample is refluxed for 2 hours in a strong acid solution of concentrated H_2SO_4 with Ag_2SO_4 (5.5 Ag_2SO_4 per kg of H_2SO_4) and with a known excess of potassium dichromate ($K_2Cr_2O_7$) (Fig.2.10.) After digestion, the remaining unreduced $K_2Cr_2O_7$ is titrated against ferrous ammonium sulphate (FAS) to determine the amount of $K_2Cr_2O_7$ consumed. The oxidizable organic matter is calculated in terms of oxygen equivalent. Also run blank containing the reagents and a volume of distilled water equal to that of the sample.

$$\text{COD},(\text{mg/L}) = \frac{(A - B) \times N \times 8000}{\text{Sample volume, mL}}$$

Where,

A = Volume of FAS used for blank

B = Volume of FAS used for sample

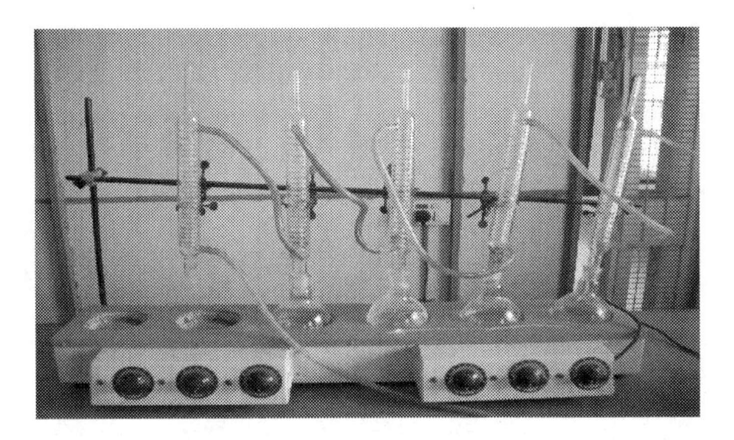

Fig.2.10. COD experiment set up

c. Estimation of total organic carbon (TOC)

The TOC was estimated by the wet digestion method of Walkley and Black as described by Piper, 1989. The diluted 20 ml sample is digested with 50-75 ml of 1N $K_2Cr_2O_7$ with 20ml of conc. H_2SO_4. After 30 minutes, 10 ml of ortho phosphoric acid is added. This was titrated against 1 N Ferrous Ammonium Sulphate (FAS) with diphenylamine as indicator. Run blank also without sample in a similar way.

$$\text{TOC}(\%) = \frac{(Bv - Sv) \times NFAS \times 100 \times 0.03}{Vs}$$

Where,

B_v = Volume of FAS used for blank in titration (mL)

S_v = Volume of FAS used for sample in titration (mL)

NFAS = Normality of FAS

Vs = Volume of test Sample

Organic matter (%) = Organic carbon (%) × 1.724

Total N (%) = Organic matter (%) × 0.05

d. Estimation of total Kjeldhal nitrogen (TKN)

Available nitrogen is estimated in the samples by micro kjeldhal method. To 1 ml of sample, 2-3 ml of 25% $KMnO_4$ solution was added followed by a few drops of Conc. H_2SO_4. To this, 10-15 ml of di-acid (H_2SO_4 & $HClO_3$ in the ratio 5:2) is added and the digestion was carried out in a Kjel plus digestion unit. 5 ml of each of the digested samples is distilled with 20 to 50 ml of 40% NaOH and the distillate titrated against 0.05 N H_2SO_4.

$$\text{Total Kjeldhal Nitrogen, mg/L} = \frac{\text{Titre value} \times 14 \times \text{Vol. of acid makeup}}{\text{Volume of acid pipetted}} \times 100$$

Chapter - 3

Biogas Technology

3.1. Introduction

Biogas is a popularly known renewable source of energy. Biogas production is based on the phenomenon of biological decomposition of organic materials in the absence of air. Biogas technology is however, something more than the mechanism of producing this useful fuel gas. Through this technology very good organic manure, improved rural sanitation and health care and more importantly prevention of deforestation and thus ecological balance can be achieved. As energy is now being looked at from a global perspective, ecology and environment has also become global concern.

On the energy front, it is already well known that the days of total dependence on fossil fuels are over and frantic efforts are being made all over the world for adoption of renewable sources of energy. Biogas, being one of the potential renewable source of energy has a definite role to play because it reconciles two apparently conflicting objectives of getting cheaper and better fuel for cooking, lighting, running engines and good organic manure to supplement and optimize the use of chemical fertilizers from cattle dung and agricultural residues. Hence, a biogas plant is an asset to a farming community. Farmers conceive the importance of biogas plant in terms of the availability of large quantities of better quality manure. One-third to half of cattle dung is burnt as fuel and thus lost to the soil. A biogas plant doubles the availability of organic manure. Besides, many social and environmental benefits are also derived from biogas production. India, having the largest cattle population in the world, offers a tremendous potential for the development of biogas.

3.2. Biogas

Biogas is produced through anaerobic digestion of organic matter, particularly human, animal, vegetable and plant wastes, in the absence of air to produce an inflammable gas which mainly consists of methane (CH_4) and carbon dioxide (CO_2). The composition and basic information of the biogas is given in Table 3.1 and 3.2.

Table 3.1. The composition of biogas

S.No	Composition of gas	Content (%)
1	Methane	55-65
2	Carbon dioxide	55-45
3	Hydrogen	Trace
4	Hydrogen sulphide	Trace
5	Ammonia	Trace

Table 3.2. Basic information about biogas

Particulars	Equivalent
1m^3 of biogas	1000 liters
1 cft of biogas	0.028 m^3
1 m^3 of biogas	35.6 cft
Gas requirement for cooking	0.3 to 0.4 m^3/day / person
Gas requirement for lighting	0.10 to 0.12 m^3 / h /100 candle power light
Gas requirement running the engine	0.4 to 0.5 m^3/ HP/ h
Calorific value of biogas	3500 to 4800 Kcal / m^3
Optimum gas to air ratio for complete combustion	1.6 to 7
Gas production per kilogram of wet dung	0.04 to 0.05 m^3
Optimum temperature for maximum gas production	30 to 35°C
Optimum cow dung water mixing ratio	1:1
Optimum pH range for biogas production	6.8 to 7.2
Optimum retention period for the cattle dung slurry	40 to 50 days

Optimum carbon to nitrogen ratio	30 to 35.1
Availability of dung with cows or bullocks	About 10 kg / day
Availability of dung with buffaloes	About 15 kg / day
Availability of dung with calves	About 5 kg / day
Methane content of biogas	55 to 65%

3.2.1. Science of biogas production

Biogas production involves basically four stages namely, hydrolysis, acidogenesis, acetogenesis and methanogenesis by the action of respective bacteria.

3.2.1.1. Stages involved in methane production

Formerly, methane fermentation was hypothetically designed as a two-step process including an acid forming stage followed by a methane forming stage. However, with the exception of methanol, acetate and formate, methanogenic bacteria cannot metabolize alcohols and organic acids. The current scheme represents a three-stage process in which three groups of bacteria are involved.

a. Hydrolytic and acidogenic bacteria

b. Acetogenic bacteria

c. Methanogenic bacteria

As it can be seen, the three groups of bacteria include a wide variety of microorganisms (Fig.3.1). This diversity reflects the great number of substrates and intermediary metabolites involved. The main impediment in this field arises from the difficulty encountered in assessing the *in vivo* importance of the different groups of bacteria.

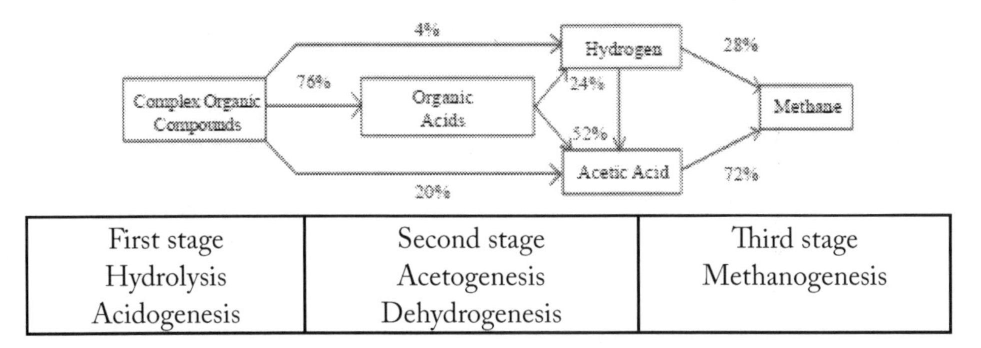

Fig. 3.1 Stages involved in methane production

a. Hydrolysis

This involves hydrolysis and liquefaction of large organic molecules by extracellular enzymes produced by obligate anaerobic bacteria such as *Clostridium*. Starch and Glycogen are hydrolyzed to disaccharides by the action of amylases.

These enzymes attack polysaccharides from the non-reducing end of the chain cleaving alternate glycosidic bonds. Subsequently the disaccharides are cleaved to monosaccharides by a glycosidase. The specific enzyme involved depends on the nature of the glycosidic bond and the monosaccharide involved i.e. whether the configuration is α or β, dextro or laevo and the size of the heterocyclic ring.

Cellulose is hydrolyzed to cellobiose and subsequently to glucose by cellulase and cellobiase which include both exo and endo glucanases. Lipases and esterase's hydrolyse fats and lipids.

When the substrate is a triglyceride of long chain fatty acids, the hydrolysis proceeds in a stepwise manner, with the rapid formation of di and mono glycerides followed by slow hydrolysis of monoglycerides. Proteases catalyse the cleavage of peptide bonds of proteins. Exopeptidases are restricted to terminal peptide bonds and endopeptidases attack peptide bonds that are centrally located in the peptide chains.

b. Acid production

After the hydrolysis the initial degradation products can be utilized by microorganisms provided they are able to pass into the cells through the plasma membrane. This membrane selectivity regulates the flux of nutrients, ions and waste products in and out of the cell. It is composed of lipids and proteins. The lipids provide the structural properties and the proteins provide the membrane with its distinctive functional properties. It is believed that the various proteins embedded in the membrane act as carriers for specific substances or types of molecules against a concentration gradient, the process known as active transport.

Degradation of Monosaccharides

A typical metabolic pathway for the degradation of monosaccharides is the Embden-Mayerhof-Pamas (EMP) pathway or glycolysis. The end product of glycolysis in yeast is ethanol where as in bacteria, it is acetic acid. Carbon dioxide is evolved in either case. Energy is stored as ATP. NAD is a coenzyme often involved in hydrogen transfer reaction, which must be regenerated with coupled reactions.

Degradation of fatty acids

It is generally agreed that the anaerobic degradation of long chain fatty acids proceeds primarily *via* β-oxidation in which two C atoms at a time are split from the chain. The sequence shows the removal of one acetate unit, which combines with reduced coenzyme A (CoA-SH). Acetic acid can then either be liberated from CoA in a subsequent reaction or the acetate can be transferred to other functional compounds. FAD (Flavin Adenine Dinucleotide) is a prosthetic group for a wide variety of enzymes (E-FAD) associated with hydrogen transfer reactions.

Decomposition of amino acids

In nature and in anaerobic biological processes, it is suspected that proteolytic species of *Clostridium* are largely responsible for the anaerobic decomposition of amino acids. They can accomplish this in one or two ways. Some amino acids may be fermented individually by pathways specific to that compound.

i.Anaerobic Digestion of glutamic acid

From a molecule of glutamic acid, a molecule of pyruvic acid and a molecule of acetic acid are formed.

ii. Stick land reaction

In this reaction pairs of amino acids are fermented. One amino acid acts as the electron donor (oxidized) while other as the electron acceptor (reduced).

c. Microbes involved in hydrolysis and acid production stages

In these two steps, both anaerobic and facultative anaerobic bacteria are involved. In the case of anaerobic bacteria oxygen is toxic. Some can tolerate low levels of oxygen (non-stringent or tolerant anaerobes) but others (stringent or strict anaerobes cannot even tolerate even low level and may die upon brief exposure to air. Facultative anaerobic bacteria grow in the presence or absence of molecular oxygen. In metabolic terms facultative anaerobic forms fall under two categories.

1. Organisms like lactic acid bacteria have an exclusively fermentative energy yielding metabolism but relatively insensitive in the presence of oxygen. Here, the organic matter turns out to be both electron acceptor and electron donor.

2. Many yeast and enteric bacteria can shift from a respiratory to a fermentative

mode of metabolism. They use oxygen as terminal electron acceptor when it is available but can also obtain energy in its absence by fermentation reaction.

There are micro-aerophilic bacteria also which grow well at a partial pressure of oxygen considerably below that of 0.2 atmosphere present in air. The fourth group of organisms which take part in the hydrolysis and acid production steps are the sulphate and nitrate respires. They in the absence of molecular oxygen use nitrates, sulfates, sulfites, thiosulphates and sulphur as electron acceptors. Meanwhile there are obligate sulphur respires also.

The genera, which have been isolated from anaerobic systems, which do not directly involve in methane production but associated with the hydrolysis and acid production, are *Clostridium, Aerobacter, Bacillus, Escherichia, Micrococcus, Paracholobacterium, Proteus, Pseudomonas, Sarcina* and *Streptomyces.* During the hydrolysis and fermentation steps, very little COD is removed, although some CO_2 is formed during the process which consists largely of breaking down large molecules into one or two carbon units. The growth yield of fermentative cells is low due to the small net production of ATP by substrate phosphorylation.

d. Methane production stage

The low molecular weight acids produced in the acid production stage are further degraded to methane and carbon dioxide by the highly specialised group of bacteria referred to as methane producing bacteria or methanogens. Methanogens are subdivided into two groups

Hydrogenotrophic methanogens

These organisms have the unique ability to couple the product of organic oxidation (hydrogen) to reduction of carbon dioxide. In this process CO_2 is the terminal hydrogen acceptor and is analogous to oxygen in aerobic respiration. They are the hydrogen utilizing chemolithotrophs, which convert hydrogen and carbon dioxide into methane. By this process they obtain energy for growth by oxidizing molecules / compounds such as hydrogen and formate. Electrons thus generated are utilized to reduce CO_2 with the formation of methane. The energy liberated during this process is used for the fixation of CO_2

$$CO_2 + 4H_2 \longrightarrow CH_4 + H_2O$$

This group helps maintain the very low levels of partial pressure of hydrogen necessary for the conversion of volatile acids and alcohol to acetate. This means that in an anaerobic digester if the partial pressure of hydrogen goes high, fatty acid synthesizing bacteria can generate long chain fatty acids such as propionate and butyrate from H_2 and CO_2.

Acetotrophic methanogens

This group of organisms ferments compounds such as acetate and methanol and form methane and carbon dioxide as the products. These are acetoclastic bacteria or acetate splitting bacteria, which derive energy by the fermentation process.

$$CH_3COOH \longrightarrow CH_4 + CO_2$$

About two third of methane is derived from acetate conversion and the one third as a result of CO_2 reduction.

3.3. Factors Affecting the Biogas Production

There are different factors which are affecting the production of biogas. The optimum conditions of the listed factors are given in the brackets.

1. pH (6.5 – 7.5)
2. Loading rate
3. Size of the material used
4. Temperature (30-35 °C)
5. Retention time (30-55 days)
6. Seeding
7. Uniform feeding
8. Diameter to depth ratio
9. Water to input material mixing ratio (1:1)
10. Total solid content of the feed materials (8-10%)
11. C:N ratio (30:1)
12. Nutrients
13. Type of feedstock
14. Toxicity due to end product
15. Acid accumulation inside the digester

3.4. Design of Biogas Plants

There are many designs and models of biogas plants in operation with each one having some special characteristics; however, each popular model has some basic components. Other than these, the fixed dome/roof model plant e.g. Janata model and the horizontal plug-flow model, have some special components for proper functioning of these designs.

3.4.1. Basic components

The basic components of plants which are common to most of the models are:

a. Digester or fermentation chamber
b. Gas holder or gas storage chamber
c. Inlet (pipe or tank)
d. Outlet (pipe or tank)
e. Mixing tank
f. Gas outlet pipe
g. Inlet and outlet displacement chambers
h. Plant dome or arched roof
i. Inlet and outlet gates
j. Manhole or digester cover

3.4.1.1. Types of designs

a. Floating gas holder digester (KVIC)

The floating gas holder digester which is used in India is known as Khadi Village Industries Commission (KVIC) plant. This type was developed in India and is usually made of masonry. It runs on a continuous basis and uses mainly cattle dung as input material. The gasholder is usually made of steel, although new materials such as ferro cement and bamboo-cement have already been introduced. The original version of this floating gasholder digester was a vertical cylinder provided with partition wall except for the small sizes of 2 and 3 m^3 of gas per day. The main characteristic of this type is the need for steel sheets and welding skill.

b. Fixed dome digester

In fixed dome digester, the gas holder and the digester are combined. This digester, which was developed and is widely used in China, runs on a continuous batch basis. Accordingly, it could digest plant waste as well as human and animal wastes. It is usually built below ground level, hence it is easier to insulate in a cold climate. The digester can be built from several materials, *e.g.* bricks, concrete, lime concrete and lime clay. This facilitates the introduction and use of local materials and manpower. The variable pressure inside the digester was found to cause no problems in China in the use of the gas.

c. Modified fixed dome digesters

In the Chinese fixed dome digester, a manhole is provided on the top of the plant. By modifying this design, two designs of Janata and Deenbandhu fixed dome digesters have been developed. In the Janata fixed dome digester, the bottom is flat concrete, the digester is cylindrical wall and the dome is a segment of sphere with the brick masonry. In the Deenbandhu fixed dome biogas digester, the lower part of concrete is segment of sphere and the upper part of brick masonry is hemisphere.

d. Biogas Plant Models (India)

The different types of family size biogas plants available at present in our country include

1. KVIC (Khadi and Village Industries Commission) design.
2. PRAD (Planning, Research and Action Division) design
3. Murugappa Chettiar Research Centre design.
4. Tamil Nadu Agricultural University dome type design
5. ASTRA (Application of Science and Technology to Rural Areas) design
6. Himachal Pradesh Capsule design
7. Kacha-Pucca model of Punjab Agricultural University
8. Plug-flow design
9. AFPRO (Action for Food Production) design

10. Roorkee design
11. Deen Bhandhu design
12. Fibreglass fixed dome design (Underground model)
13. Mobile biogas plants
14. Plastic emulsion coated, heavily insulated, temperature controlled Switzerland biogas plants.
15. IARI (Indian Agricultural Research Institute) design
16. Ganesh Model
17. Ferro-cement Digester Biogas Plant

The various designs listed above have their own merits and demerits. They are popular in different regions depending upon the choice of the beneficiaries. Evaluation of all the above designs at a particular research station over a period of time may be a worthwhile proposition in order to arrive at their suitability for extension.

3.5. Construction of Biogas Plants

3.5.1. General considerations

Proper size (capacity) of the plant should be fixed based on the cattle strength and family size. Proper place should be selected for the installation of biogas plant based on the location of cattle shed, kitchen, drinking well, manure pit, elevation of the area, sunshine and water table of the site. Concrete mix and cement mortar should be prepared on clean and smooth surface. The richness and thickness of concrete mix and cement mortar should be based on the water table and type of soil. First class (quality) bricks should be used for the construction of bio-gas plant. Grooves should be made at the bricks joint so that inner plastering can be given easily.

The best quality gas holder as per the specification should be purchased from Government approved workshop. The bottom of the mixing tank should be slightly higher than the top of the digester. The top level of the outlet pipe should be at least 0.45 m below the top of the digester. During the construction of digester wall, bricks should be laid in header layer and stretcher layer alternatively, so that joint will not be in the straight line. Outer side of the digester wall should be back filled and properly compacted with sand.

3.5.2. KVIC digester

It mainly consists of two main parts: (1) Digester or pit (2) The gas holder or the gas collector. The drawings of the KVIC biogas plant of capacity up to 3 and 4 cu.m and above are given in Fig. 3.2.

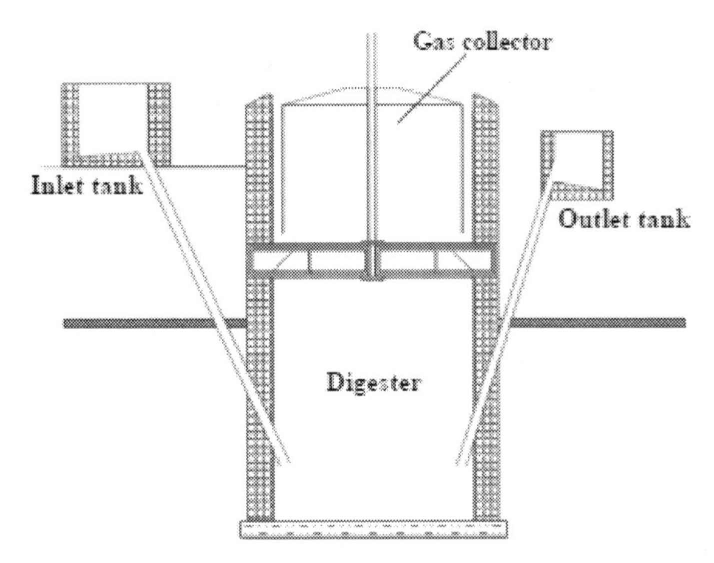

Fig. 3.2. KVIC biogas plant

3.5.2.1. Digester

Also called as the fermentation plant, it is a sort of well of masonry work, dug and built below the ground level. The depth of this well varies from 3.5 to 6 metres, and diameter from 1.35 to 6 metres, depending upon the gas generating capacity and the quantity of raw material fed each day. The digested well is divided vertically into two semi-cylindrical compartments by means of partition wall in the centre. The partition wall is lower than the level of the digester rim and hence it is submerged in slurry when the digester is full. Two slanting cement pipes reach the bottom of the well on either side of the partition wall. One pipe serves as the inlet and the other as outlet. An inlet chamber near the digester at surface level serves for mixing dung and water which is done mechanically or manually. The mixture of dung and water in proportion of 4:5 by volume, called slurry, flows down the inlet pipe to the bottom of the primary compartment of the digester. The digester is designed to the raw material for 60 days.

This ensures enough stay time of the input material for complete digestion.

The outlet chamber is again at surface level, just a few cms below the level of the inlet chamber. If both compartments of the digester are full and if more slurry is added from the inlet, then an equivalent amount of fermented slurry flows out of the outlet and discharged into the compost pit.

3.5.2.2. Gas holder

It is a drum constructed of mild steel sheets, cylindrical in shape with a conical top and radial support at the bottom. It fits into the digester like a stopper. It sinks into the slurry due to its own weight and rests upon the ring constructed for this purpose. As the gas is generated the holder rises and floats freely on the surface of the slurry. A pipe is provided at the top of the holder for flow of gas for usage. To prevent the holder from tilting a central guide pipe is fitted to the frame and is fixed to the bottom in the mansonary work.

The pressure, under which the gas is generated in this arrangement, varies between 7-9 cm of water column. The holder also acts as a seal for the gas. The cost of the holder constitutes almost 40% of the total cost of the digester.

Since the gas must be dried before it is sent to the over for use, it is better to send the gas through a vessel filled with soda lime. This will help in better utilization of available calorific value. Without much pressure drop and difficulty, the gas can be sent to 15 to 30 m distances from the source point. To ensure leak proof joints the ends of the metal part are lined in such a way as to prevent shocks. Generally, the pit is deep and narrow, but at places where the water level is low, the design has been modified and the volume has been taken horizontally.

The floating drum is metallic and it is given an anti-corrosive paint coating. If not property maintained, the drum corrodes soon and the life of the plant very much gets reduced. Perhaps the cost and the maintenance factors of this type of digesters are prohibitive factors for being not very much liked by users. However, the construction is quite simple and the gas comes out at the constant pressure.

The only maintenance required is the painting of the gas collector at regular intervals. Many other materials like ferrocement, fibre, glass, reinforced polymer gas holders are being tried instead of M.S., but all of them are quite expensive. To avoid corrosion of the gas holder, a jacketed top of the pit is also suggested wherein the jacket some type of non-corrosive oil is filled once for all.

3.5.3. Deenbandhu biogas plant

The Deenbandhu model biogas plant has been developed by making use of the principles of structural engineering as well as the long experience of AFPRO of working with fixed-dome biogas plants. The advent of Deenbandhu biogas plant is mainly to reduce the initial cost of installation as well as the subsequent maintenance cost. The design essentially consists of segments of two spheres of different diameters joined at their bases. The structure thus formed acts as the digester and the gas storage chamber, and also provides for empty sphere over the contents of the digester. The higher compressive strength of the brick masonry and concrete makes it preferable to go in for a structure which could be always kept under compression. A spherical structure loaded form the convex side will be under compression and therefore, the internal load will not have any residual effect on the structure. The design and construction of Deenbandhu biogas plant is simple (Fig.3.3).

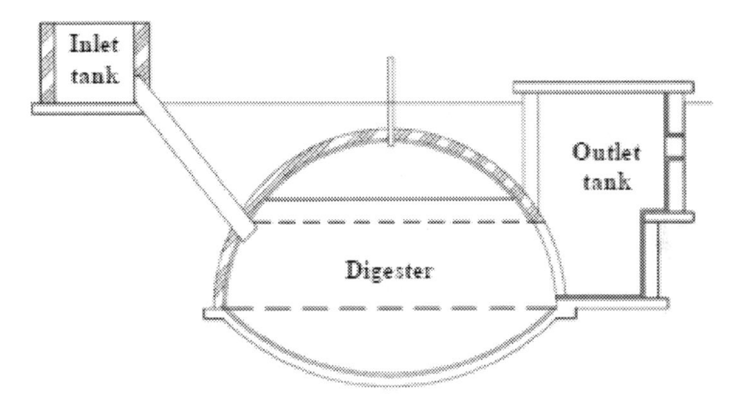

Fig. 3.3. Deenbandu biogas plant

3.6. Comparison between Fixed Dome and Floating Drum Models

a. Merits of Fixed-Dome type

Floating gas holder type		Fixed-dome type
(i)	Capital investment is high	Capital investment in the corresponding size of biogas unit is low
(ii)	Steel gas holder is a must which needs to be replaced after few years due to corrosion damage	Steel gas holder is not required

(iii)	Cost of maintenance is high		As there is no moving part, the maintenance cost is minimized
(iv)	Life span of the digester is expected to be 30 years and that of gas holder is 5 to 8 years		Life span of the unit is expected to be comparatively more
(v)	Movable drum does not allow the use of space for other purposes		As the unit is an underground structure, the space above the plant can be used for other purposes.
(vi)	Effect of low temperature during winter is more		Effect of low temperature will be less
(vii)	It is suitable for processing of dung and night-soil slurry. Other organic materials will clog the inlet pipe		It can be easily adapted / modified for use of other materials along with dung slurry
b. Merits of Floating Gas-Holding Type			
(i)	Release of gas is at constant pressure		Release of gas is at variable pressure which may cause slight reduction in the efficiency of gas appliances. To operate a diesel engine, attachment of a gas pressure regulator in the pipeline is a must.
(ii)	Construction of digester is known to masons but fabrication of gas holder requires workshop facility		Construction of the dome portion of the unit is a skilled job and requires thorough training of masons.
(iii)	Location of defects in the gas holder and repairing are easy		Location of defects in the dome and repairing are difficult
(iv)	Requires relatively less excavation work		Requires more excavation work
(v)	In areas having a high water table, horizontal plants could be installed.		Construction of the plant is difficult in high water table areas

3.7. After Construction of the Plant

3.7.1. General

1. Initial filling of the slurry should be done after 2 weeks of curing of the plant. Care should be taken to avoid the entry of sand and gravel with the slurry in to the plant.

2. The initial gas obtained should be let out because it will contain more of carbon dioxide. The right type of burner designed for biogas (as per ISI specification) should be used for efficient utilization of gas. Do not open the gas valve before lighting the gas appliances.

3. While boiling water or cooking food stuff, regulate the gas valve in such a way to utilize more gas during the first half and less gas during the second half (ie. after boiling) in order to reduce gas consumption. Gas regulator cock or valve of the burner should be tightly closed.

4. If there is any gas leakage in the burner, doors and windows of the kitchen should be opened to allow the fresh air to enter and the gas to escape. Never allow build-up of gas pressure in the dome/gas holder. Do not keep the gate valve loosely to avoid the wastage of the gas.

5. Fill the plant with properly mixed dung slurry. (mixing ratio of dung: water =1:1) Recommended quantity of slurry mixed (i,e. 25 kg. dung with 25 litres of water per cubic meter plant) should be added daily.

6. Remove the condensed water from pipe line at regular intervals.

7. The burner, burner holes and jets should be periodically cleaned with kerosene and small iron wire to avoid blocking of gas. Wide bottom utensils should be preferably used so that loss in heat will be minimum. Pressure cooker should be used to reduce the gas consumption.

8. Adjust the flame by turning the gas cock and air regulator till blue flame is obtained. Do not test the biogas flammability at the top of the biogas plant.

9. Do not allow scum to form in the digester, otherwise the production of gas will be reduced. Once the manure pit gets filled up with ejected slurry from biogas plant, it should be removed to the field or compost pit.

3.7.2. Designs

i. KVIC biogas plant (Floating drum)

When charging fresh slurry in the bigger sized floating drum type plants, make sure that it is filled equally on both sides of the central partition wall. After the initial filling of the digester tank, drum moves upwards due to gas pressure. Regular loading of the plant should be commenced after this.

Rotate the gas holder once or twice every day in order to break the scum. Once in a year the gas holder should be painted with oil paint or black paint to avoid rusting.

ii. Deenbandhu biogas plant

During the initial filling, fill the plant with slurry up to second step level of the outlet (bottom of the gas chamber). The regular loading of the plant should be commenced only after automatic ejection of the slurry through the outlet opening. Proper loading of the plant will avoid the scum formation because of the slurry movement. The entire biogas plant should be covered with soil to a minimum thickness of 15 cm.

3.8. Applications of Biogas

Biogas and digested slurry has multiple applications for the benefit of individual households as well as the whole of the village and nation. The utilization of biogas and byproducts are illustrated in Fig.3.4.

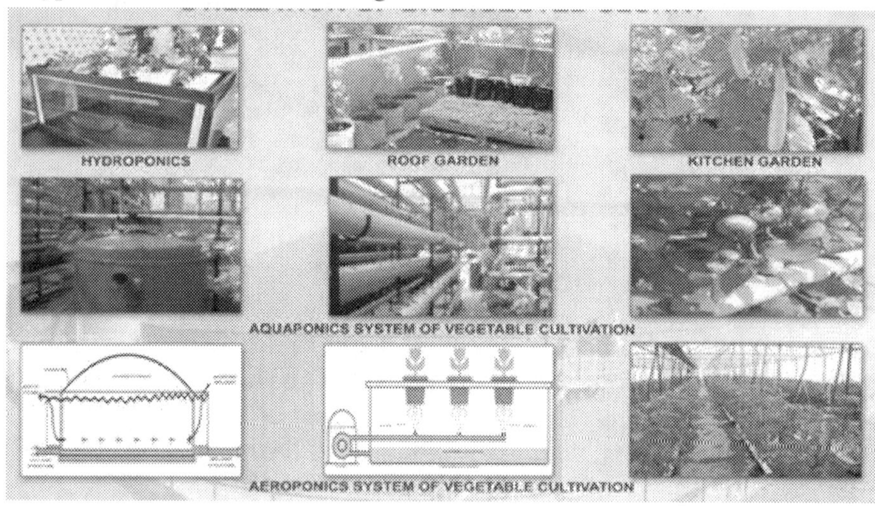

Fig.3.4. Utilization of biogas and byproducts

Some of the important applications are:

- Cooking fuel
- Fuel for lighting
- Fuel for motive power to replace diesel oil
- Enriched organic manure of agriculture and aquaculture

- Manure for mushroom growing
- Manure for seed coating
- Use of light to trap insects at the farm
- To treat human excreta and for pollution control

Another important aspect of biogas technology is in relation to improving soil fertility. It is well known that physical structure of the soil is as important as its nutrient contents. Biogas plant manure, in addition to providing macro elements like N, P and K and trace elements like iron, copper, boron etc., also improves the soil water retention capacity because of its humus content. Apart from this, the biogas plant manure buffers pH change in the soil and thus equilibrium of soil is maintained.

3.8.1. Biodigested Slurry (BDS)

Biodigested slurry is the by-product obtained from the biogas plant after the digestion of the dung and generation of the fuel gas. The biodigested slurry is very good manure. It is very similar to the farm yard manure available in the farm but it becomes very much different after the digestion process. This, biodigested slurry becomes free of weed seeds, odour etc and so it is a highly valuable organic manure than the farmyard manure. The elemental analysis of biodigested slurry by X-ray reveals that silicon, phosphorous, sulphur, potassium, calcium and iron are present with 65.595, 8.275, 3.117, 7.623, 11.693 and 3.698% respectively. Among the other elements, increased amount of Silicon also play an important role in the plant growth as well in the yield.

a. Biodigested slurry as organic manure

While supplementing the biodigested slurry, it has to be kept in mind that each plant has a specific N, P, K requirement and it also varies from place to place according to the soil nature. Generally, 50% substitution has given best results. Hence, the amount of chemical fertilizer can be reduced to half and as we know the average N, P, K contents of biodigested slurry, we can clearly find out the quantity need to be supplemented. Therefore, partial or full replacement of chemical fertilizer can reduce the problem of soil infertility. In addition, it reduce pollution, save energy required to produce chemical fertilizer, provide biogas as an alternative energy source and finally provide a better life and good yield. The biodigested slurry is used in hydroponics, aeroponics and aquoponics.

Hydroponics is a subset of hydroculture, which is the growing of plants in a soil less medium, or an aquatic based environment. Hydroponic growing uses mineral nutrient (Biodigested slurry) solutions to feed the plants in water, without soil.

Aeroponics is the process of growing plants in an air or mist environment without the use of soil or an aggregate medium. Unlike hydroponics, which uses a liquid nutrient solution as a growing medium and essential minerals to sustain plant growth; or aquaponics which uses water and fish waste, aeroponics is conducted without a growing medium. It is sometimes considered a type of hydroponics, since water is used in aeroponics to transmit nutrients to plants by mixing biodigested slurry in water.

Aquaponics is essentially the combination of Aquaculture and Hydroponics. Water is pumped up from the fish tank into the growbed. The water trickles down through the media, past the roots of the plants before draining back into the fish tank. The plants extract the water and nutrients they need to grow, cleaning the water for the fish. Biodigested slurry acts as a food to fish and nutrient to plants.

b. Biodigested slurry and soil properties

Biodigested slurry besides supplying plant food, benefits the soil by influencing several physical properties of the soil. It helps in soil aggregation and increasing the soil porosity. It enhances the water holding capacity, prevents leaching of nutrients, buffers pH changes and improves several other properties like resistance to soil erosion. Forms complexes with harmful elements like copper, aluminium and minimize plant toxicity. It absorbs herbicides, pesticides and prevents their washing. It supplies N, P and S during degradation. It can be used in any type of soil and for any crop. The application of recommended dose of manure about 10 tonnes per ha in irrigated areas and 5 tonnes per ha in dryland areas, increases the yield by 10 - 20% of fruit trees, tea, coffee, sugarcane, millets, rice, jute, vegetables like carrots, bhendi, radish and sweet potato.

c. Biogas as an insecticide for stored seeds

Biogas containing 60% methane and 40% carbon dioxide has been tested as a fumigant for insect control in certain pulses. The biogas fumigation has killed all the pests in about 10-12 days at a pressure of 1.4 kg/cm^2. The biogas fumigation did not affect seed germination, seeding vigour or grain quality and no gaseous methane residue were left in the gain.

d. Biogas slurry as seedling root dipping medium for rice

Studies have been conducted to evaluate the effect of freshly bio-digested slurry as seedling root dipping medium for rice. The results showed that seedling root dipping with a combination of bio-digested slurry, *Azospirillum*, Phosphobacteria, Zinc sulphate and Phytohormone enhanced the growth and yield attributes. Root dipping of seedling with this combination of nutrients increased the grain yield by 5.9% over no dipping. Uptake of N, P and K also enhanced by various root dipping. Uptake of Zinc was not influenced by dipping treatments.

e. Biogas slurry as seed dressing agency on seed performance

Seeds of maize and rice were soaked in digested slurry and water for 12 h, 24 h and 48 h. The slurry treated seeds were also surface washed with water to remove any adherents from the seed surface. Treated seeds were evaluated in the laboratory (28°C, 95% RH) dark and in soil filled pots for germination percentage, moon germination time and seedling growth; Application of slurry significantly improved the seed performance over untreated seeds but its effect was equal or less pronounced than water treatment. Removal of slurry adherents from seed surface before germination caused improvement in above parameters.

3.8.2. Biogas plant slurry applications

The following are the different methods of applying biodigested slurry as manure.

a. Air dried biogas slurry can be applied by spreading on the agricultural land at least one week before sowing the seeds or transplanting the seedlings.

b. The liquid slurry can be mixed directly with the running water in irrigation canal which will enable spreading of the slurry uniformly in the cropped area or in cultivation land.

c. Biogas slurry can also be coated on the seeds prior to sowing. This acts as insecticide and prevents seeds or plants from insect attack. This helps in early germination and healthy growth of seedlings.

d. The digested slurry is fed through the channel, flowing over a layer of green or dry leaves and filtered in the bed. The water from the slurry filters down which can be reused for preparing fresh dung slurry. The semi-solid slurry can be transported easily as it was in the consistency of fresh dung and used for top dressing of crops like sugarcane and potato.

e. Biodigested slurry is also being used for fish culture, which acts as a supplementary feed. On an average, 15-25 litres of wet slurry can be applied per day in a 1200 sq. pond. Slurry mixed with oil cake or rice- bran in the 2:1 ratio increases the fish production remarkably. In general organic manures about 10 t / ha, in the form of FYM or compost or biodigested slurry is recommended to be applied once in three years to maintain the organic content of soil, besides providing nitrogen, phosphorous and potassium in the form of organic fertilizers to the crop.

Chapter - 4

Liquid Biofuels

Depleted supplies of fossil fuel, regular price hikes of gasoline and environmental damage have necessitated the search for economic and eco-friendly alternative to gasoline. By 2020, India's dependency on oil imports is projected to go beyond 90% from 80% at current level. So it is time to shift for alternate fuels like bio-diesel, bio-ethanol and bio-butanol which is produced from biomass which is a renewable energy source and abundantly available.

4.1. Bio-diesel

Bio-diesel is a variety of ester-based oxygenated fuels derived from natural, renewable biological sources such as vegetable oils. The name indicates, use of this fuel in diesel engine alternate to diesel fuel. Bio-diesel operates in compression ignition engines like petroleum diesel thereby requiring no essential engine modifications. Moreover, it can maintain the payload capacity and range of conventional diesel. Bio-diesel fuel can be made from new or used vegetable oils and animal fats. Unlike fossil diesel, pure bio-diesel is biodegradable, nontoxic and essentially free of sulphur and aromatics.

4.1.1. Importance of Bio-diesel

Bio-diesel is monoalkyl ester of long chain fatty acids produced from the transesterification reaction of vegetable oil with alcohol in the presence of catalyst and can be used as fuel. Bio-diesel is a fuel produced through the transesterification reaction between a triglyceride and an alcohol, typically methanol or ethanol using sodium hydroxide as catalyst. This fuel has several advantages as compared to the petroleum-based diesel. The main reason for its utilization is that the bio-diesel has lower environmental impact than the diesel fuel obtained from petroleum. Bio-diesel is produced from vegetable oils or fats that are renewable resources, and hence no accumulation of carbon dioxide in the atmosphere. The amount of sulphur is around 11 ppm. In a regular diesel sulphur content is 1600 ppm, while in low sulphur diesel it is 500 ppm and for ultra-lower sulphur diesel it is 50 ppm. The cost associated with this level of purification is too high and the alternative of using bio-diesel as an additive became very attractive.

Additionally, bio-diesel presents a high lubricity that extends the engine life and bio-diesel is not toxic and easily biodegradable. The production of bio-diesel using material of high acidity and therefore lower value requires the development of a process combining different types of catalysis such as acids and alkalis. The acidity of these materials is given by free fatty acids which in the presence of a base and water are transformed into soaps.

4.1.2. Chemistry of bio-diesel production

Bio-diesel is produced by transesterification of large, branched triglycerides in to smaller, straight chain molecules of methyl esters, using an alkali or acid or enzyme as catalyst. There are three stepwise reactions with intermediate formation of diglycerides and monoglycerides resulting in the production of three moles of methyl esters and one mole of glycerol from triglycerides. Alcohols such as methanol, ethanol, propanol, butanol and amyl alcohol are used in the transesterification process. Methanol and ethanol are used most frequently, especially methanol because of its low cost, and physical and chemical advantages. They can quickly react with triglycerides and sodium hydroxide is easily dissolved in these alcohols. Stoichiometric molar ratio of alcohol to triglycerides required for transesterification reaction is 3:1. In practice, the ratio needs to be higher to drive the equilibrium to a maximum ester yield.

$$\text{Triglyceride} + \text{ROH} \xleftrightarrow{\text{catalysts}} \text{Diglyceride} + \text{R}^{\text{I}}\,\text{COOR}$$

$$\text{Diglyceride} + \text{ROH} \xleftrightarrow{\text{catalysts}} \text{Monoglyceride} + \text{R}^{\text{II}}\,\text{COOR}$$

$$\text{Monoglyceride} + \text{ROH} \xleftrightarrow{\text{catalysts}} \text{Glycerol} + \text{R}^{\text{III}}\,\text{COOR}$$

Transesterification has been classified into four major types. They are,

- Alkali catalyzed transesterification
- Acid catalyzed pretreatment

- » Catalyst free transesterification
- » Enzyme catalyzed transesterification

4.1.3. The *Jatropha curcas* plant and oil

The oil yielding plant *Jatropha curcas* L. is a multipurpose and drought resistant large shrub, which is widely cultivated in the tropics as a live fence. The Jatropha plant can reach a height up to 5 m and its seed yield ranges from 7.5 to 12 tonnes per hectare per year, after five years of growth. The oil content of whole Jatropha seed is 30-35% by weight basis. The problems in use of jatropha oil as fuel in diesel are listed in Table 4.1.

Table 4.1. Problems in use of jatropha oil as fuel in diesel engine

Problems	Causes
Chocking of injectors on piston and head of engine	High viscosity of raw oil, incomplete combustion of fuel. Poor combustion at part load with raw oil
Carbon deposits on piston and head of engine	High viscosity of oil, incomplete combustion of fuel.
Excessive engine wear	High viscosity of raw oil, incomplete combustion of fuel. Dilution of engine lubricating oil due to blow by of raw oil

The above problems can be solved by converting raw jatropha oil into bio-diesel through transesterification process.

4.1.4. TNAU Bio-diesel pilot plant

The bio-diesel pilot plant consists of a transesterification reactor with heater, a stirrer, chemical mixing tank, glycerol settling tank (2 No.) and washing tank. The capacity of pilot bio-diesel plant is 250 litres/day. The cost of the pilot plant is Rs. 5 lakhs.

4.1.4.1 Pilot bio-diesel plant operation

In the pilot bio-diesel plant, jatropha oil is mixed with alcohol and catalyst mixture in transesterification reactor. The reactor is kept at reaction temperature for specific duration with vigorous agitation. After reaction, the bio-diesel and glycerol mixture is sent to the glycerol settling tank. The process flow chart for bio-diesel production from *Jatropha curcus* seeds is shown in Fig.4.1.

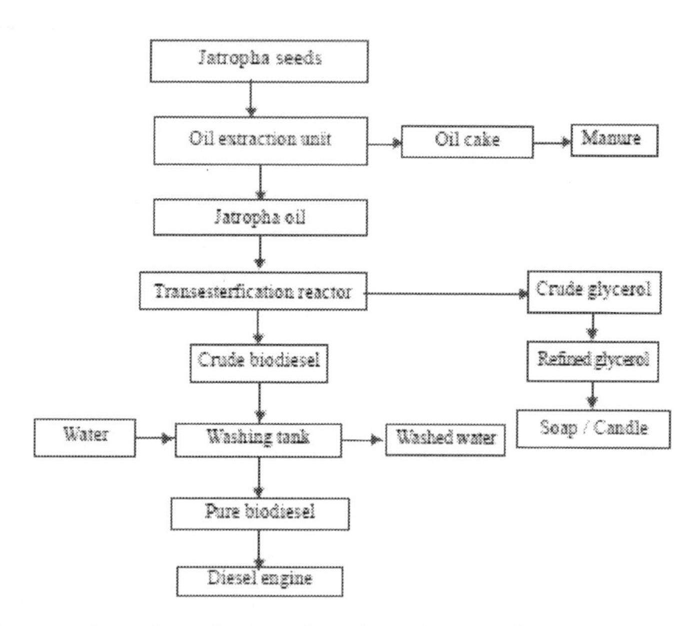

Fig.4.1. Process flow chart for bio-diesel production from *Jatropha curcus* seeds

The crude bio-diesel is collected and washed to get pure bio-diesel. Depending upon the need, the size of the unit can be scaled up to get higher production.

9.1.5. Properties of Bio-diesel

a. Viscosity

Viscosity is a measure of the ability of a liquid to flow or a measure of its resistance to flow. It is also defined as the force required to move a plane surface of area 1 square meter over another parallel plane surface 1 meter away at a rate of 1 meter per second when both surfaces are immersed in the fluid; the higher the viscosity, the slower the liquid flows.

b. Viscosity index

An arbitrary scale used to show the magnitude of viscosity changes in lubricating oils with changes in temperature.

c. Flash point

It is the minimum temperature at which an oil/petro fuel gives out sufficient vapor to form an inflammable mixture with air and catches-fire momentarily i.E., Flashes, when flame is applied. Bio-diesel has a flash point (150°c) well above the flash point of petroleum based diesel fuel (±70°c). Flash point less than 23 °c is dangerous and highly inflammable. The safe value is greater than 60-66 °c.

d. Fire point

It is the lowest temperature at which vapors given off by oil ignites and continues to burn for at least 5 seconds. In most cases, fire point is 5-40 °c higher than flash point. It gives an idea of fire hazards during the storage and use of oil.

e. Cloud point

When oil is cooled at a specified rate, the temperature at which it becomes cloudy or hazy is called the cloud point of oil. This haziness is due to separation of crystals of wax or increase of viscosity at low temperature. It is determined by cold filter plugging point test.

f. Pour point

The temperature at which the oil ceases to flow (or pour) is called the pour point. Lubricants used in a machine working at low temperature should possess low pour points; otherwise solidification of lubricant will cause jamming of the machine. Presences of waxes in the lubricating oil raise the pour point.

g. Freezing point

It is the temperature at which the fuel oil freezes completely and cannot flow at all. This is important in case of aviation gasoline because at higher altitudes where lower temperatures are encountered, the fuel supply from fuel tank to engine may be impeded due to choking of pipeline if the freezing point of the fuel oil is not sufficiently low. Aviation gasoline should have freezing point below 60 °c to avoid trouble due to crystal formation in feed lines and filters.

h. Octane number

This is property of gasoline which is used in a spark ignition engine and expresses its knocking characteristics. Higher the octane number better is the fuel as maximum permissible power increases with the octane number besides causing minimum knocking. Octane number of hydrocarbon fuels increases in the order of n-paraffins, olefins, naphthenes, isoparaffins and aromatics.

i. Cetane number

Cetane number (cn) is a dimensionless descriptor of the ignition quality of a diesel fuel (df). It is property of diesel and is used to indicate its quality and performance in compression ignition engine. Cetane ($c_{16}h_{34}$) has a very small ignition delay period hence it is given a cetane number rating 100. A- methyl naphthalene has a very high ignition delay period; hence it is given a rating of zero. Regular diesel fuel has a cetane number of 42-45 which varies with feed stock.

Cetane number increases in the order aromatics, iso-paraffins, naphthene, olefins, n-parafins, whereas octane number decrease in the same order. The oil of high cetane number has a low octane number and vice versa. High cetane number diesel fuel will facilitate easy starting of c.I. Engines particularly in cold weather, faster than warm up, increased engine efficiency and power output and reduced exhaust smoke, odour and combustion noise.

j. Oxidation stability

Oils tend to oxidize in contact with air rapidly at higher temperature. Oxidation products are insoluble, gummy and sticky materials which may clog oil holes, filters, pipelines and other parts of the oil circulation system. Soluble oxidation products are corrosive. Hence resistance to oxidation is a desirable property of oil. Oxidation inhibitors added to improve their resistance. It gets oxidized first and the oil is protected.

k. Acid value

Acid value or number determines the acidity of oil. It is defined as the number of milligrams of the koh required to neutralize the free acid present in 1 gm of oil.

l. Colour and fluoresence

Colour of petroleum products indicates the degree of refinement. Distillates are lighter in colour than the residual oils. A pale colour does not indicate lower viscosity. The mineral oils show fluorescence in reflected light whereas fixed oils do not show fluorescence. Hence, fluorescence helps to detect adulteration.

m. Sulphur content

Sulphur content increases with the increase in the boiling range of the products. Sulphur content in oil is determined by chemical method. Sulphur is largely present in oil in the form of corrosive organic compounds. On combustion, it gives off foul gases corrosive in nature.

The fuel properties of jatropha bio-diesel produced in the pilot plant are given in the table 4.2.

Table 4.2. Fuel properties of Jatropha oil and its bio-diesel

Properties	Jatropha Oil	Jatropha bio-diesel	Diesel
Density, g/ml	0.920	0.865	0.841
Viscosity @ 40°C, Cst	3.5	5.2	4.5
Calorific value, MJ/kg	39.7	39.2	42.0
Flash point, °C	240	175	50
Cloud point, °C	16	13	9

4.1.6. Alternate feedstock for bio-diesel production

In India, there are several non-edible oil seed species apart from *Jatropha curcas* (Jatropha), such as *Pongamia pinnata* (Karanja), *Azadirachta indica* (Neem), *adhuca indica* (Mahua) etc., which could be utilized as a source for production of oil and can be grown in large scale on non-cropped marginal lands and waste lands.

4.1.7. Byproducts and utilization

Bio-diesel production produces three major bio-diesel waste products. They are glycerin, methanol and waste water

4.1.7.1. Glycerin

For every 10 parts of bio-diesel produced, there is 1 part of glycerin is formed.

The utilization of Glycerin is as follows:

- Glycerin can be converted into ethanol through fermentation with a strain of E. coli. This method of producing ethanol actually has higher yields than the traditional way of producing ethanol.

- Researchers have found that when glycerin is combined with a harmless strain of E.coli, succinate is formed. Succinate is used to flavor food and drinks, is an ingredient in perfumes and dyes, and in nontoxic plastic solvents. Succinate has a large market in the US.

- » Formate can be produced from glycerin and E-coli strains. It is a common preservative in animal feed, and is also used as an antibacterial agent.
- » Glycerin can be used to produce a hydrogen rich gas for use in fertilizers, food production and chemical plants.
- » It can also be taken to a wastewater treatment plant where it will be added to a methane digester.

4.1.7.2. Bio-diesel waste

Bio-diesel waste can be composted by mixing in organic matter such as leaves, grass clippings, straw, and twigs. Toss the mixture around every couple weeks to oxygenate it in order to help it decompose. This will also allow any remnant methanol to evaporate.

4.1.7.3. Remnant methanol

Excess methanol can be boiled off the glycerin within a closed container and directed into a condenser. It can then be reused in the next batch of bio-diesel.

4.1.7.4. Waste water

The source of waste water is the water used to wash the bio-diesel. The waste water can be treated and reused.

4.1.8. Advantages of bio-diesel

1. Produced from sustainable / renewable biological sources
2. Ecofriendly and oxygenated fuel
3. Sulphur free, less CO, HC, particulate matter and aromatic compounds emissions
4. Income to rural community
5. Fuel properties similar to the conventional fuel
6. Used in existing unmodified diesel engines
7. Reduce expenditure on oil imports
8. Non toxic, biodegradable and safety to handle

4.2. Bio-ethanol

Alcohol possesses most of the desirable attributes of a source of energy. Ethanol or ethyl alcohol is commonly known as alcohol which is one of the most important and popular industrial fermented products. It is a primary alcohol (single -OH group), colourless, volatile and miscible in water in all proportions.

4.2.1. Raw materials for bio-ethanol production

Ethanol can be produced in sufficient quantities from a wide range of raw materials. These raw materials include sugar, starch, lignocellulosic materials, such as vegetables, growing crops, waste organic matter such as straw and sawdust, tropical grasses, farm waste, molasses, water gas, industrial waste like sulphite liquor from paper and pulp industry, etc.

Environmental and energy security concerns are forcing us to shift to alternate energy. Currently, biofuel production is minimal and to enhance its production we require policy support, community and local interest, technological breakthrough and cost effective feedstock production.

4.2.2. Properties of ethanol

All alcohols have a common feature. Their molecular structure includes a hydroxyl (OH) radical, which gives them certain characteristics, high latent heat of vaporization and solubility in water. This water like characteristics is most apparent in the alcohols of low molecular weight, ethanol and methanol, because the OH radical predominates over their short hydrocarbon chains. Of the innumerable family of substances known as the alcohols, only the first two members of the monohydric saturated series, namely, methanol and ethanol, the most volatile alcohols, are of interest as motor fuel. For the purpose of comparison, the corresponding properties of gasoline a typical hydrocarbon fuel are also given in the Table 4.3.

Table 4.3. Properties of ethanol, methanol and gasoline (*H.B. Mathur, 1988*)

S.No	Property	Methanol	Ethanol	Gasoline
1.	Formula	CH_3OH	CH_3CH_2OH	C_4-C_{12}
2.	C/H mass	3.0	4.0	5.33
3.	Relative molecular mass	32.042	46.068	114.224
4.	Carbon, % by wt.	37.5	52.0	84.0

5.	Hydrogen, % by wt.	12.6	13.2	16.0
6.	Oxygen, % by wt.	49.9	34.7	0.0
7.	Density at 20°C, Kg/M^3	792.0	789.5	702.0
8.	Boiling point,°C	64.96	78.32	125.7
9.	Freezing point,°C	-94.9	-117.6	-57.0
10.	Heat of vaporization, KJ/Kg	1101.1	841.5	297.3
11.	Heat of fusion, KJ/Kg	100.5	104.7	180.0
12.	Critical temperature,°C	240.2	243.3	296.2
13.	Critical pressure, bar	97.08	61.78	24.22
14.	Critical density, Kg/m^3	358.0	280.0	233.0
15.	Compression factor at critical parameter	0.222	0.248	0.274
16.	Coefficient of cubical expansion at 20° C/K	0.00119	0.00110	0.00114
17.	Viscosity at 20°C, Ns/M^2	0.000584	0.00120	0.00054
18.	Flash point,°C (Pensky-Martin instrument)	-1.1	12.80	-43
19.	Spontaneous ignition temp.,°C	480	440	425-510
20.	Thermal conductivity at 20°C,W/mK	0.212	0.182	0.147
21.	Vapour pressure at 38°C, bar	0.320	0.210	0.800
22.	Refractive index-D (D=589.3 mu at 20°C)	1.329	1.361	-
23.	Gross heating value, KJ/Kg	22680	29770	47850
24.	Net heating value, KJ/Kg	19930	26750	44380
25.	Heating value of std. stoichiometric mixture, KJ/m^3	3253	3528	3540
26.	Stoichiometric air/fuel ratio (mass)	6.46	8.99	14.55
27.	Stoichiometric product/ reactant ratio (molar)	1.061	1.065	1.057
28.	Flammability limits:			
	a. Air/Fuel(% volume)	6.7-36.5	4.0-13.7	1.4-7.3

	b. Air/Fuel(% mass)	1.6-121.6	2.7-14.0	6.0-22.0
29.	Flame speed, cm/s	55	52	37.5
30.	Adiabatic flame temp.,°K	2229	2214	2260
31.	Octane quality			
	a. Research octane number	110	106	80-95
	b. Motor octane number	92	89	70-85
32.	Heat capacity at 25°C, J/mol K	81.6	111.5	239.3
33.	Entropy at 25°C, J/mol K	126.8	160.7	328.0
34.	Gibbs energy of formation at 25°C, KJ/mol	- 166.2	- 174.8	6.90
35.	Standard enthalpy of formation at 25°C, KJ/mol	- 238.57	- 277.63	- 259.28
36.	Standard enthalpy of reaction (net) at 25°C, KJ/mol	-638.6	-1234.9	-5069.3
37.	Dangerous concentrations for prolonged exposure, ppm	200	1000	500

Table 4.3 reveals that the alcohols are pure substances, boiling at one temperature, just as does water or lead, while gasoline has a range of boiling point and infact, contains many different substances, generally known as hydrocarbons. Secondly the gasoline contains no oxygen, alcohols contains oxygen and hence alcohol have more volatility. Volatility sets a limit to the usefulness of the alcohols as fuels. Because of the presence of substantial quantities of chemically combined oxygen in alcohols, they not only have lower calorific value but also lower stoichiometric air requirements. For complete combustion of 1 kg of gasoline, about 15 kg of air is required, for ethanol only 9 kg of air is required for each 1 kg. The internal total energy available from a fuel depends also upon the final volume of the products of combustion. Alcohols have higher octane rating which enables higher compression operation without engine knock. Moreover, alcohols are liquid state fuels that fit into the present infrastructure of the fuel distribution and sales network.

4.2.3. Ethanol production

Ethanol can be commercially produced by a variety of methods, which can be broadly classified into two categories: (a) Chemical synthesis and (b) Enzymatic fermentation. Chemical synthesis make use of stored materials, such as coal and petroleum, hence are not of lasting utility in view of the non-renewable nature of these raw materials and their fast depleting reserves. Enzymatic fermentation, on the other hand, make use of naturally occurring raw materials which are renewable in nature.

Any material which is capable of being fermented by enzymes can serve as a source for ethanol production. These raw materials can be categorized into three groups: (i) sugar containing materials, such as sugarcane, sugar beet, sweet sorghum and molasses - a byproduct of sugar industry, etc., (ii) starch containing materials, such as corn, algae, cassava etc. and (iii) cellulosic materials, such as bagasse, biomass, wood wastes, agricultural and forest residues, agro industrial products or wastes like sulphite liquor from paper and pulp industry, sawdust from saw mills etc.

There are two key reactions to understand how biomass is converted to bio-ethanol,

- **Hydrolysis** is the chemical reaction that converts the complex polysaccharides in the raw feedstock to simple sugars. In the biomass-to-bio-ethanol process, acids and enzymes are used to catalyze this reaction.
- **Fermentation** is a series of chemical reactions that convert sugars to ethanol. The fermentation reaction is caused by yeast or bacteria, which feed on the sugars.

$$\underset{\text{Glucose}}{C_6H_{12}O_6} \rightarrow \underset{\text{Ethanol}}{2\,CH_3CH_2OH} + \underset{\text{Carbon dioxide}}{2\,CO_2}$$

4.2.3.1. Ethanol production from sugar materials

Sugar containing materials provide the most potential feedstock for ethanol production. Sugars (saccharides) are low molecular weight carbohydrates composed of one or more saccharose group(s). The simplest are monosaccharides. A hexose is a monosaccharide composed of six carbon atoms. Glucose and fructose are two hexoses capable of being fermented. Although they have the same formula - $C_6H_{12}O_6$ - their molecular structure is different. In more complex sugars, the monomeric units (disaccharides and polysaccharides) are joined together by a chemical bond called the glycolytic link and change the complex sugars into monosaccharides (fermentable sugars). The ethanol yield from different sugar materials are listed in Table 4.4. The flowchart of ethanol production from sugar materials is shown in Fig.4.2.

Table 4.4. Ethanol yield per1000 kg feed stock from different sugar materials

S.No	Sugar crop	Total fermentable solids (kg)	Ethanol yield (kg)
1.	Sugarcane	112.0	54.3
2.	Sorghum	96.0	46.0
3.	Sugarbeet	66.7	32.0

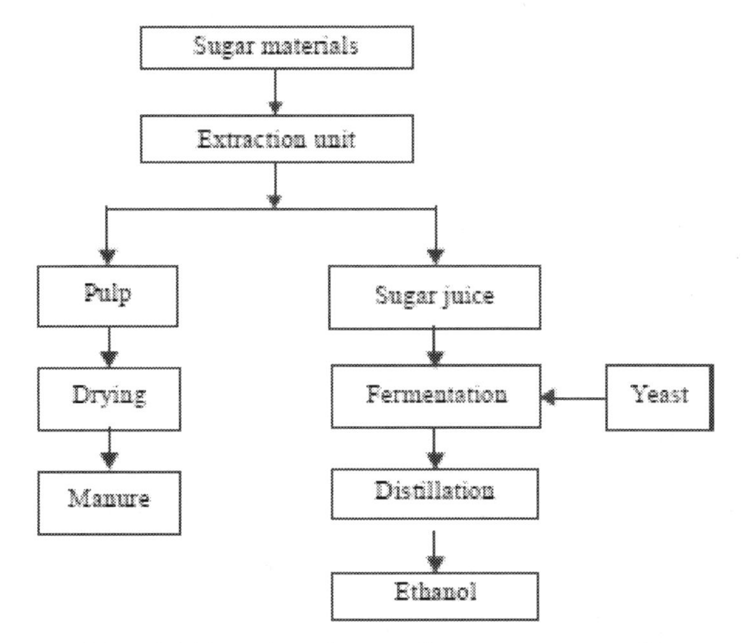

Fig.4.2. Ethanol production from sugar materials

4.2.3.2. Ethanol from starch materials

Starch based feedstock include a variety of cereals, grains and tubers like Cassava. Starchy feedstock can be hydrolyzed to get fermentable sugar syrup and give an average yield of upto 42 litres of ethanol per 100 kg of feedstock. The flowchart of ethanol production from starch based materials is shown in Fig.4.3.

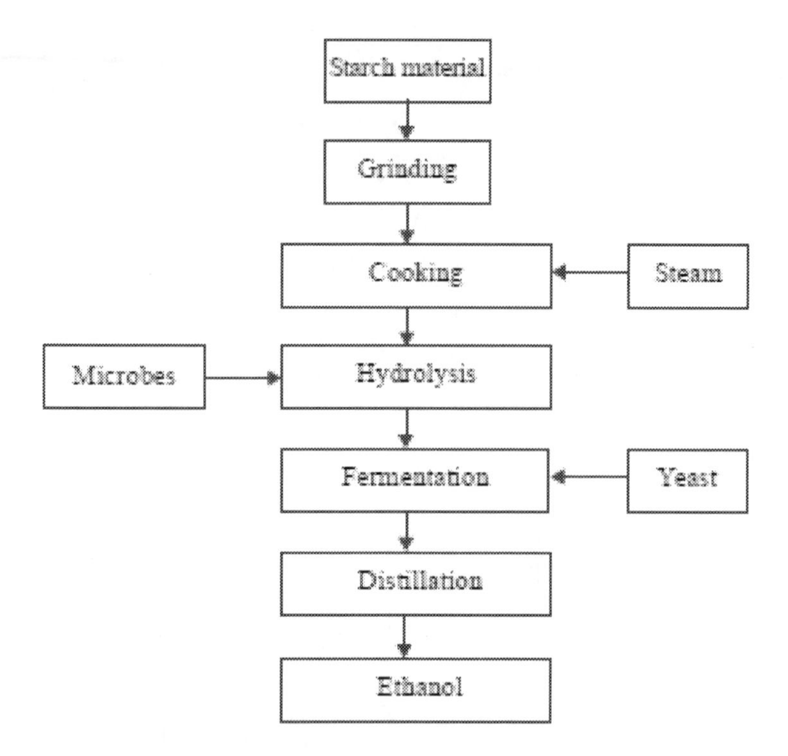

Fig.4.3. Ethanol production from starch based materials

4.2.3.3. Ethanol production from cellulosic materials

Cellulosic ethanol i.e. ethanol from forestry or agricultural waste is considered a way to prevent displacement of edible crops. The process of producing ethanol from sugarcane, corn, cassava and other starch materials will be simple and yields more ethanol as compared to cellulosic materials. And sugar and starchy materials are used as food materials, so ethanol production from these materials may affect the supply of food products. The flowchart of ethanol production from cellulosic materials is shown in Fig.4.4.

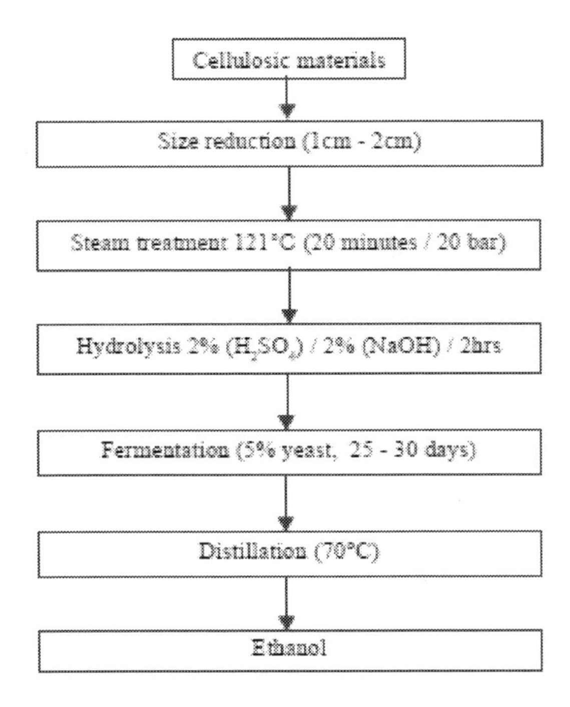

Fig.4.4. Ethanol production from cellulosic materials

Ethanol from cellulosic materials which is mainly a waste is a better way and also efficient in waste utilization. Sugarcane bagasse, wheat straw, wood, and effluent from paper & pulp industries are the very potential cellulosic feed stocks.

4.2.3. Fermentation types

Although fermentation is only one step in the production process, it is the key step. Fermentation is a biochemical process in which enzymes produced by microorganisms transform an organic substance (substrate) into ethanol. The substrate used in ethanol fermentation is a six-carbon sugar. The microorganisms used in fermentation whether yeasts, molds or bacteria - contain no chlorophyll and therefore cannot produce their own food through photosynthesis. Rather, they produce enzymes that act as catalysis to convert carbohydrates from organic material into energy. These enzymes start the fermentation process and the biological changes that produce ethanol from the sugar substrate. Yeasts are the most commonly used microorganisms responsible for producing the enzymes that convert glucose to ethanol. The complex reactions and interrelationships of yeast fermentation of a sugar substrate is called the Embden-Meyerhoff parnas pathway. Yeast species used in ethanol fermentation are unicellular fungi which exhibit a number of different strategies for growth and multiplication. The organisms of primary interest include - *Saccharomyces cerevisiae, S.uvarum, Schizosaccharomyces. pombe.*

4.2.3.1. Batch Fermentation

The sugar solution supplemented with yeast nutrients is added to the fermenter and fermenter is inoculated with a rapidly growing culture of yeast from the seed tank. Usually the time required to completely utilize the substrate is 36-48 h. The overall productivity from this process is about 1.8 - 2.5 kg of ethanol produced per m^3 fermenter volume per hour. After completion of fermentation the cells are removed from the broth medium before distillation.

Advantages

1. Control of microbial contaminants
2. Control of ethanol quality per batch

Disadvantages

1. Higher capital requirement for large scale production because of the number of fermenters required.
2. Decrease in volumetric efficiency of fermenters, based on time used.
3. Possible variations from batch to batch, requiring homogenization.

4.2.3.2. Continuous fermentation

In the continuous process, sugars and nutrient mediums are continuously added to the reactor. The fermenter consists of a vertical cylindrical tower with a conical bottom. The tower is topped by a large diameter yeast settling zone fitted with baffles. The sugar solution is pumped into the base of the tower which contains a plug of flocculent yeast. Reaction proceeds progressively as the beer rises, but yeast tends to settle back and be retained. High cell densities of 50 to 80 g/h are achieved without an auxiliary mechanical separator. Residence times of below 4 hours have been possible with sugar concentration upto 12% sucrose giving 90% conversion of ethanol. This process is called continuous tower process.

Advantages

1. Higher volumetric efficiency

2. Yeast recycling
3. Establishment of flow growth-rate equilibrium
4. A more consistent product than can be obtained from batch fermentation.
5. Lower capital and labour cost
6. Reduced time requirements

Disadvantages

1. Higher potential for serious microbial contamination
2. Inspite of providing a more consistent product, homogenization is still required.

4.2.3.3. Fed batch Fermentation

This fermentation is intermediate of both batch and continuous fermentation. Sterile nutrients are added in increments. The Characters of fed batch fermentation are

» Initial medium concentration is relatively low.

» Medium constituents are added continuously or in increments.

» Controlled feed results in higher biomass and product yields.

» Fermentation is still limited by accumulation of toxic end products.

» Products are harvested in one stroke.

4.2.3.4. Solid state fermentation

The growth of microorganisms is on moist solid substrate particles in the absence or mere visible liquid water between the particles. The moisture content of solid substrate ranges between 12-80%. The water content of a typical submerged fermentation is more than 95%. Solid state fermentation is usually used for the fermentation of agricultural products or foods, such as rice, wheat, barley, corn and soybeans.

4.2.3.5. Separate Hydrolysis and Fermentation (SHF)

In order to break down the hemicellulose and cellulose to sugars, the basic structure of the biomass must be attacked. Once the structure of the biomass is disrupted, the hemicellulose and cellulose can be converted to sugars enzymatically. This can be done by the use of acid known as acid hydrolysis or by enzymes known as enzymatic hydrolysis.

4.2.3.6. Simultaneous Saccharification and Fermentation (SSF)

It is a combination of acid hydrolysis, enzymatic hydrolysis and fermentation. Here, the cellulose feedstock is pre-treated with dilute acid pretreatment. The process is often run as a steam explosion treatment of acid impregnated material, where the hemicelluloses are removed and the digestibility of the cellulose is enhanced. The choice of enzymes preparation used in SSF is subject of great importance due to very high cost of the enzymes. However, presently a combination of cellulase and *Saccharomyces cerevisae* (yeast) is used. Cellulose catalyzes the saccharification while the yeast is used to produce ethanol.

Simultaneous saccharification and fermentation (SSF) is one process option for production of ethanol from lignocellulosic materials. The principal benefits of performing the enzymatic hydrolysis together with the fermentation, instead of in a separate step after the hydrolysis, are the reduced end-product inhibition of the enzymatic hydrolysis, and the reduced investment costs. The principal drawbacks, on the other hand, are the need to find favourable conditions (*e.g.* temperature and pH) for both the enzymatic hydrolysis and the fermentation and the difficulty to recycle the fermenting organism and the enzymes. To satisfy the first requirement, the temperature is normally kept below 37°C, whereas the difficulty to recycle the yeast makes it beneficial to operate with a low yeast concentration and at a high solid loading. Significant progress has been made with respect to increasing the substrate loading, decreasing the yeast concentration and co-fermentation of both hexoses and pentoses during SSF.

Inevitably, there are also disadvantages of SSF in comparison to the separate hydrolysis and fermentation (SHF) process. The optimum temperature for enzymatic hydrolysis is typically higher than that of fermentation at least when using yeast as the fermenting organism. In an SHF process, the temperature for the enzymatic hydrolysis can be optimized independently from the fermentation temperature, whereas a compromise must be found in an SSF process.

Furthermore, the yeast cannot be reused in an SSF process (Fig.4.5) due to the problems of separating the yeast from the lignin after fermentation. The enzymes are equally difficult to reuse, also in an SHF process.

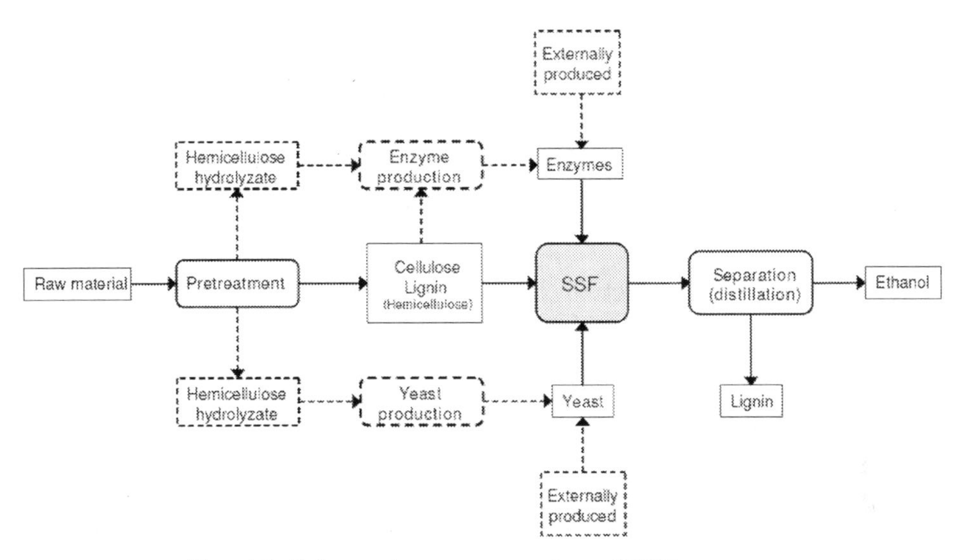

Fig.4.5. Schematic representation of SSF process

Apart from SSF and SHF, the alternatives are consolidated bio processing (CBP) and simultaneous saccharification and cofermentation (SSCF). In CBP, enzyme production, biomass hydrolysis and fermentation occur in a single reactor.

4.2.5. Types of fermentors

Conventional method - Three major types based on the presence or absence of oxygen and requirement of stirring.

i. Non stirred non aerated reactors - Used for production of traditional products. e.g.: wine, beer, cheese etc.

ii. Non stirred aerated reactors - Used much rarely, fermentation using Fungi

iii. Stirred and aerated reactors - Most often used for production of metabolites which require growth of microbes which require oxygen. Most of the newer methods are based on this type of bioreactors.

4.2.5.1. Stirred tank bioreactor

The conventional mixing reactor is made up of either glass or stainless steel. Stirrer can be either at the top or bottom of the reactor. The dimensions of the reactor depend on the amount of heat to be removed from the vessel. Baffles in the centre of the tank prevent formation of vortex and effective mixing of the ingredients.

Advantages

1. Lower investment needs
2. Lower operating costs

Disadvantages

1. Foaming is often a problem, can be overcome using proper antifoaming agents.
2. Some antifoaming agents inhibit the growth of microbes.

4.2.5.2. Air lift bioreactors

The stirred tank bioreactors lack well defined flow of air. Compressed air pumped from below is used for aeration and agitation is based on the draft tube principle. It eliminates the need for a stirrer system / mechanical agitation.

Advantages

1. Lower friction

1. Lesser energy requirements
2. The mechanical parts are easy to construct
3. Lower operating costs

Disadvantages

1. Capital needed is more
2. Difficulty of sterilization
3. Efficiency of mixing is low

4.2.6. Uses of ethanol

a. Solvent and industrial chemical

Ethanol is one of the largest volume organic chemicals used in industry where its solvent properties are employed in a wide range of products and processes. The quantity of ethanol used as a solvent rank second only to the amount produced as potable spirits. Alcohol is sold for solvent purposes in products such as intracellulose coatings, shellacs, varnishes, inks, hydraulic fluids, liquid detergents and soaps, deodorants, perfumes, antiseptics and lotions. It is also used by the cosmetic, pharmaceutical and food industries in the production of vitamins, flavours and essences, mouthwashes, blood products and fortified wines.

b. Chemical feedstock

In most developed countries, the synthetic organic chemicals industry relies very largely on petroleum as its source of feedstock. Ethanol, produced itself synthetically from ethylene, is used in the production of glycol ethers, ethyl amines, ethylvinyl ether and ethyl esters but has been replaced in recent years by other petroleum-derived substrates in the production of acetaldehyde, acetic acid, butadiene, butyraldehyde and ethylhexanol. The uses of ethyl alcohol are listed in Table 4.5.

Table 4.5. Uses of ethyl alcohol

S.No.	General use	Specific use
1.	Fuel	Light
		Power
		Heat
2.	Raw material in chemical processes	Ethylene
		Ether
		Esters
3.	General utility	Hospital
		Chemical laboratory
		Home
4.	Solvent	Dyes
		Nitrocellulose
		Oils and waxes
		Drugs and chemical
		Preparation of tinctures
	Miscellaneous	Carbon remover
		Preservative
		Antiseptic

c) Fuel

Ethanol is a liquid fuel, which can be produced from renewable resources and can be alternative to automotive fuel. The technology of using alcohol with gasoline in spark ignition engines has been developed. A mixture of 90% gasoline and 10% ethanol (Gasohol) is being marketed in some countries. In India, National Policy on biofuels has an indicative target of 20% blending of biofuels, both for bio-diesel and bio-ethanol, by 2017 is proposed. However, the chief deterrents to the wide-spread use of ethanol as a fuel have been economics and the availability of adequate quantity of alcohol.

4.2.7. By-products in ethanol fermentation

a. Waste Biomass

Due to the anaerobic nature of ethanol fermentation, the overall synthesis of biomass is limited. In general, a 10% substrate feed with 95% conversion to alcohol will yield 5.0 g/l of dried cell mass. For recycle processes, a return of 35-40% total biomass in the broth is all that is required to meet fermentation demands. After concentration, microbial biomass may be dried and utilized as a high protein food or feed supplement.

b. Stillage

Stillage is the waste water produced after ethanol has been removed by distillation from a fermented mash. For every volume of ethanol produced by a distillery, about 15 times that volume of stillage is produced. This effluent has an extremely higher Biological Oxygen Demand and its discharge, untreated, into natural water courses can cause severe environmental problems.

c. Fusel oil

Fusel oil is the mixture of volatile, pungent-tasting alcohol produced during alcoholic fermentation. Although they are considerably less volatile than ethanol, a simple distillation with steam (the analyser column) usually results in their complete carry-over into the distillate, but rectification to produce a high proof product allows the separation of a fusel oil fraction.

d. Carbon dioxide

For every cubic meter of ethanol formed about 760 kg of CO_2 is liberated from the fermentation system. Of this total, 70-80% can be recovered in a closed system.

After purification to remove aldehydes and alcohol, the gas may be stored in cylinders or further compressed to solid or liquid form. In the gaseous state, it may be used to carbonate soda beverages or to enhance the agricultural productivity of greenhouse plants. Liquid CO_2 is frequently used in fire extinguishers, refrigeration processes and as a feed stock in the chemical industry. The primary use of solid CO_2 is as a refrigerant. It is compressed to between 17.5 and 21 kg/cm^2, cooled, condensed and passed to bulk storage containers. It can be converted to solid carbon dioxide in hydraulic presses if required.

4.3. Bio-butanol

Butanol is produced from fossil fuels called petro-butanol and from agricultural waste called bio-butanoal. Bio-butanol and petro-butanol has the same chemical properties. Advanced biofuel and conventional petroleum based fuels, such first generation biofuels is relatively low carbon emissions, and there are environmental benefits such as reduced sulfur. Butanol is considered as a potential next generation biofuel.

Butanol has 30% more energy than ethanol, which can be used in petrol engine which is designed for cars to run with gasoline blended with 85%. Where 85% blended ethanol cannot be used in petrol engine without modifications.

The process is the fermentation of biomass with Weizmann organism (*Clostridium bacterium*) which yields butanol, acetone and ethanol. Bio-butanol fermentation anaerobic process is similar to the yeast fermentation of sugars to produce ethanol.

Butanol was tradionally produced by ABE fermentation - the anaerobic conversion of carbohydrates by strains of *Clostridium* into acetone, butanol and ethanol. Acetone–butanol–ethanol (ABE) fermentation is a process that uses bacterial fermentation to produce acetone, n-Butanol, and ethanol from carbohydrates such as starch and glucose. The process of the 3 parts of acetone, butanol and ethanol to 1 part produces 6 parts. Bio-butanol fermentation feedstock such as straw and corn stalks, sugar beets, sugar cane, corn, grain, wheat and cassava, as well as non-edible energy crops such as energy crops in agricultural products. Fermentation feedstock for bio-butanol and ethanol are the same.

Butanol from agricultural waste products can be more efficient than ethanol and methanol production. The octane rating of gasoline is almost equal to butanol and does not contribute to engine knocking. Best butanol tolerates water contamination and therefore it is very suitable for ethanol fuel distribution through existing pipelines is less corrosive.

4.3.1. Advantages of bio-butanol

1. Biobutanol can be directly used in its pure form or can be blended in any concentration with gasoline, while bioethanol can only be blended up to 85% or used in its pure form in specially designed engines.

2. Regardless of the use of biobutanol as a clean vehicle fuel or gasoline, there is no need to make any modifications to the existing car.

3. It is safer to handle because it has a lower vapor pressure than bioethanol.

4. It can be mixed with gasoline at the refinery before storage and distribution because it is not hygroscopic, while bioethanol can only be mixed with gasoline just before use.

5. Because biobutanol is immiscible with water, it is less likely to contaminate groundwater if spilled; while bioethanol is completely miscible with water and will cause water pollution if spilled.

6. Unlike bioethanol, biobutanol is less corrosive, so it can be used in infrastructure such as pipelines, tanks, gas stations, pumps, etc.

7. Due to its higher energy content, it has a higher mileage/gasoline ratio

8. Compared to bioethanol, biobutanol has more similarities in the amount of calorific value, octane number and air-fuel ratio with real gasoline, which means that biobutanol is more similar in properties to gasoline.

4.4.2. Disadvantages of bio-butanol

1. The production of biobutanol is relatively low. The production rate of biobutanol from ABE fermentation is 10-30 times lower than that of bioethanol produced from the yeast ethanol fermentation process.

2. Although biobutanol has a higher energy density than other low-carbon alcoholic biofuels, its calorific value is still lower than that of real gasoline or diesel, so the fuel flow needs to be increased when used as a fossil fuel substitute.

3. Biobutanol is a type of alcohol-based fuel, so it is still not compatible with some fuel system components and can cause gas gauge reading errors in vehicles with capacitive fuel level gauges.

4. Biobutanol can produce more greenhouse gas emissions per unit of motive power extracted compared to bioethanol because it contains a lower octane number. Higher octane means higher compression ratio and efficiency, and higher engine efficiency can result in lower greenhouse gas emissions.

5. The higher viscosity of biobutanol may lead to a potential corrosion or aggrega- tion problem when used in gasoline engines.

Chapter - 5

Briquetting Technology

Recently, biomass energy is an attractive energy source because the accumulation of carbon dioxide in the atmosphere for the production and use of biomass is zero. Agricultural residues are one of the most promising biomass energy options. In our country, it is mostly a free and domestic source of energy. The main disadvantage of using energy from biomass is its bulky nature, energy. The production of agricultural residues is estimated at 650-700 million tons/year. In that whole amount 350 million tonnes of residues are surplus which is burnt in the open field. These residues can yield huge amount of energy containing fuel if it is used appropriately. The problem of handling and transport of these residues due to low bulk density can be rectified by briquetting technology. This process increases the fuel values of the residues.

The compaction of biomass in the form of solid fuels has yielded a simple solution of complex problems related with low density, high volume and expensive handling, transportation and storage of most of the biomass. The compacted fuel is more or less similar to coal and has potential to replace the conventional solid fuels and even diesel to meet the local energy needs of various sectors. These briquettes burns as good as coal, leave less ash content, emit less or no smoke and have low ignition point.

5.1. Biomass Materials Suitable for Densification

Almost all agro residues can be briquetted. Agro residues such as sawdust, rice husk, tapioca waste, groundnut shell, cotton stalks, pigeon pea stalks, soybean stalks, coconut pith, mustard stalks, sugarcane bagasse, wood chips, tamarind pods, cowpea husks, coffee husks, dried tapioca stick, coconut shell powder are commonly used raw materials for briquetting in India. Factors that mainly affect the choice of raw materials are moisture content, ash content, flow properties, flow properties, particle size and availability in the locality. Moisture content in the range of 10 to 15% is preferred, as higher moisture content will cause grinding problems and more energy is required for drying. The ash content in the biomass affects its slag behavior together with the operating conditions and the mineral composition of the ash. Biomass raw material with an ash content of up to 4% is advantageous for densification. Granular homogeneous materials that flow easily in conveyors, bunkers and storage silos are suitable for densification.

5.2. Densification

Biomass densification may be defined as compression or compaction to remove inter and intra-particle voids. This technology can help in expanding the use of biomass in energy production, since densification improves the volumetric calorific value of a fuel, reduces the cost of transport and can help in improving the fuel situation in rural areas. Five densification processes are practiced commercially namely, baling, pelleting, cubing, briquetting and extrusion.

a. **Compression baling and roll-compressing process** can reduce biomass volume to one-fifth its loose bulk. These processes are useful for agricultural biomass and certain types of forest biomass. e.g. crops from energy plantations, logging residues.

b. **Pelleting** employs a hard steel die which is perforated with frequently placed holes 0.3 to 1.3 cm in diameter. The die rotates against inner pressure rollers, forcing the continued biomass through the holes with pressures of 7 kg/mm^2. As the biomass is extruded through the die, small dense pellets are broken off at a specified length.

c. **Cubing** is a modification of pelleting which produces larger cylinders or cubes, 2.5 to 5.0 cm in diameter.

d. **Briquetting** is defined as the compaction of biomass feedstock between rollers with cavities, producing forms like charcoal briquettes.

e. **Extrusion** uses a screw extruder to force a feedstock under high pressure into a die thereby forming large cylinders (2.5 to 10 cm) of densified biomass. Binding agents such as paraffin or resin are often added to modify properties of the fuel logs

5.2.1. Advantages

1. The process increases the net calorific value of material per unit volume
2. The end product is easy to transport and store
3. The fuel produce is uniform in size and quality
4. It helps solve the problem of residue disposal
5. It helps to reduce deforestation by providing a substitute for fuel wood.
6. The process eliminates the possibility of spontaneous combustion waste
7. The process reduces biodegradation of residues

5.2.2. Disadvantages

1. High investment cost and energy input to the process
2. Undesirable combustion characteristics such as poor ignitability, smoking, etc are often observed.
3. Tendency of briquettes to loosen when exposed to water or even high humidity weather

5.3. Densification Process

There are four basic steps involved in the densification process namely, collection of raw materials, preparation of raw materials, compaction and cooling and storage.

5.3.1. Collection of raw materials

In general, any material that will burn but is not the right shape, size or form to be easily used as a fuel is a good candidate for briquetting.

5.3.2. Preparation of raw materials

The preparation of raw materials includes drying, size reduction, mixing raw materials in the right ratio, mixing raw materials with binder, etc.

a. Drying

The raw materials are available in a higher moisture content than that required for briquetting. Drying can be done in the sun, in solar dryers with a heater or hot air.

b. Size reduction

The raw material is first reduced by crushing, chopping, grinding, breaking, rolling, hammering, milling, grinding, cutting, etc., until it reaches a suitable small and uniform size (1 to 10 mm). Some materials available in the 1 to 10 mm size range do not need to be sized down. Because the downsizing process consumes a lot of energy, so it should be as short as possible.

c. Raw material mixing

It is desirable to make briquettes from more than one raw material. Mixing will be done in the right ratio so that the product has good compaction and high calorific value.

5.3.3. Compaction

The compaction process takes place inside the briquetting machine. The process depends on the briquetting technology used.

5.3.3.1. Binding agent

The binder is necessary to prevent the compressed material from springing back and possibly returning to its original form. This agent can either be added to the process or, when the woody material is pressed, it can be part of the material itself in the form of lignin. Lignin or sulfur lignin is a component of most agricultural residues. It can be defined as a thermoplastic polymer that begins to soften at temperatures above 100°C and flows at higher temperatures. The softening of lignin and its subsequent cooling under the pressure of the material is a key factor in high-pressure briquetting.

5.3.3.2. Densification techniques

Densification essentially involves two processes, compacting the bulk material under pressure to reduce its volume and agglomerating the material so that the product remains in a compressed state. Based on compaction, briquetting technologies can be divided into

1. High pressure compaction
2. Medium pressure compaction with a heating device

3. Low pressure compaction with a binder

At present two main high pressure technologies: ram or piston press and screw extrusion machines, are used for briquetting.

a. Piston press densification

A reciprocating piston pushes the material into a tapered die where it is compacted and adheres against the material remaining in the die from the previous stroke (Fig.5.1). A controlled expansion and cooling of the continuous briquette is allowed in a section following the actual die. The briquette leaving this section is still relatively warm and fragile and needs a further length of cooling track before it can be broken into pieces of the desired length. The most common type of briquette press features a cylindrical piston and die with a diameter ranging from 20-125 mm. The die tapers somewhat towards the middle and then increases again before the end. The exact form of the taper varies between machines and biomass feedstock and is a key factor in determining the functioning of the process and the resulting briquette quality.

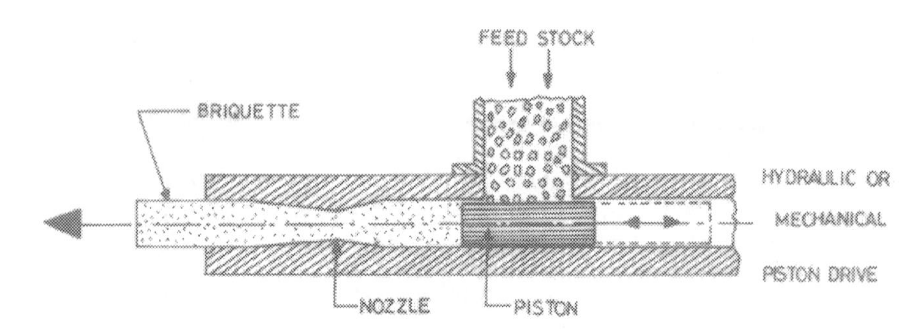

Fig.5.1. Piston press densification

b. Screw press densification

In screw press technology, the biomass is continuously extruded by a screw through a cone die that is externally heated to reduce friction (Fig. 5.2). The material is continuously fed to a screw which forces the material into a cylindrical die which is often heated to 250-300°C to raise the temperature to the point where the lignin begins to flow and the pressure increases steadily. Briquettes produced by screw presses are often of higher quality than piston press units, but the power consumption per ton of briquettes produced is also high.

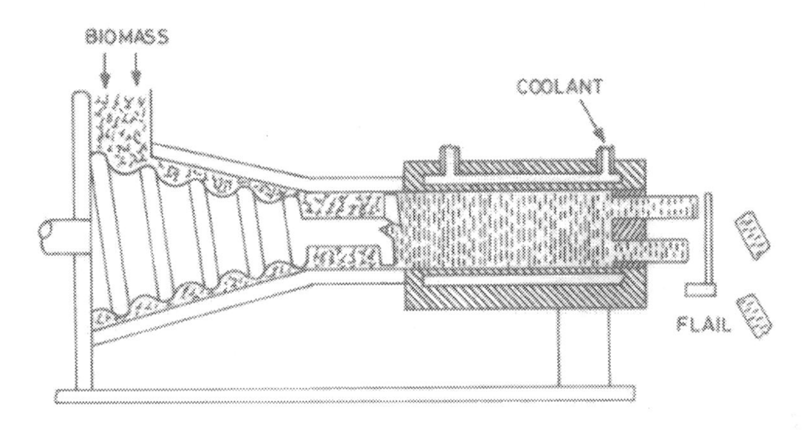

Fig.5.2. Conical screw press densification

Table 5.1. Comparison of piston press and screw press densification

Parameters	**Piston press**	**Screw Press**
Optimum moisture content of raw material (%)	10-15	8-9
Wear of contact parts	low in case of ram and die	high in case of screw
Output from the machine	In strokes	Continuous
Power consumption(kWh/ton)	50	60
Density of briquette(gm/cm^3)	1-1.2	1-1.4
Maintenance	High	Low
Combustion performance of briquettes	not so good	very good
Carbonization to charcoal	not possible	makes good charcoal
Suitability in gasifiers	not suitable	Suitable
Homogeneity of briquettes	non-homogeneous	Homogeneous

c. Pelletization

Pelleting is closely related to briquetting, except that it uses smaller dies (approximately 30 mm), so the smaller products are called pellets. The pelletizer has a series of dies arranged as holes drilled in a thick steel disc or ring, and the material is pressed into the dies by means of two or three rollers. The material to be pelletized is fed in the cylinder and when the rollers ride over this material and rotate, they push the material through holes in the die against resistance from pellets already formed in the die holes.

The two main types of pellet presses are flat and ring. The flat type is characterized by a circular perforated disk on which two or more cylinders rotate. The ring press has a rotating perforated ring on which the rollers are pressed against the inner circumference. Large capacity pelletizers are available in the range of 200 kg/h to 8 tons/h. Thus, pellet press capacity is not restricted by the density of the raw material as in the case of piston or screw presses. Power consumption falls within the range of 15-40 kWh/ton.

5.3.4. Cooling and Storage

Briquettes extruding out of the machines are hot with temperatures exceeding 200 °C. They have to be cooled and stored. Compacting biomass waste into briquettes reduces the volume by 10 times, making it much easier to store and transport than loose biomass waste. Other fuel types tend to be difficult to handle and are hazardous. Briquettes can be produced in a variety of sizes and have a long shelf-life.

5.4. Briquette Analysis

5.4.1. Determination of combustion properties

The combustion properties include percentage of volatile matter, fixed carbon, ash content and heating value. The percentage of volatile matter, fixed carbon and ash contents of four representative samples were determined based on ASTM Standard E711-87 (2004).

5.4.2. Determination of bulk density

The volume and the weight of a briquette have to be measured in order to be able to determine the bulk density. Weight measurements were performed with a normal laboratory balance, the volume of the samples was determined by using a graduated cylinder. The average bulk density was calculated from three measurement series per sample.

5.4.3. Determination of water content

The water content was measured by weighing a fuel sample (about 100 g (w.b.)) before and after drying at 105 °C according to O¨ NORM G 1074.

5.4.4. Measurement of strength and durability

The effectiveness of the briquettes created during the densification process has been measured in terms of strength and durability. Procedures used for measuring the compressive resistance, abrasive resistance, impact resistance, and water resistance of the densified products are discussed below.

a. Water resistance

Short-term exposure to rain or high humidity conditions during transportation and storage could adversely affect the quality of the densified products. Lindley and Vossoughi measured the water resistance as the percentage of water absorbed by a briquette, when immersed in water. Each briquette was immersed in water at 27 °C for one min.

b. Impact resistance

Impact resistance (or drop resistance or shattering resistance) test may simulate the forces encountered during emptying of densified products from trucks onto ground, or from chutes into bins. The drop resistance test to determine the durability of biomass pellets and briquettes. Pellets were dropped from 1.85 m height onto a metal plate 4 times. The weight retained as the percentage of the initial weight was taken as the pellet/briquette durability.

c. Water stability test

When the briquette is immersed in water, when it breaks into pieces within 5 minutes, which usually represents very low quality briquettes. When the briquette breaks into pieces under 15 minutes, it shows medium quality briquettes, and in less than 20 minutes, it is a sign of good quality briquettes. However, this method is only informative.

5.5. Utilization of briquetted fuel

Briquettes made from biomass can be used easily anywhere in place of conventional fuels like coal, fuel wood, kerosene and diesel. Effectively they can be a substitute for such non-renewable fuels without loss of efficiency.

5.5.1. Replacement for conventional fuel

Wood, coal, dung, cake and kerosene are at present most widely used fuels for domestic cooking. However, scarcity in oil supplies is likely to continue in spite of the rising cost and more judicious use. In the coal sector, the increase in production to the stipulated demand would require stupendous efforts and a high level of efficiency on the part of agencies like Coal India Ltd. The coal reserves are being exhausted as faster rate. Therefore, in village areas where wood, coal, kerosene is used as fuel, briquetted fuel will serve as a better replacement. For domestic cook stoves 12-25 mm diameter briquettes are more suitable.

Briquettes can be used in boilers, gasifiers, furnaces for thermal applications. For these applications 50-60 mm size briquettes are preferred. Briquettes of 60-90 mm size are being used in big boilers and gasifiers. Producer gas obtained from briquette based gasifiers after cooling and cleaning of tar and dust can be used for running 100% gas operated electric generating sets. On an average 1 kg biomass briquette can produce 1 kWh of electricity.

Chapter - 6

Thermochemical Conversion Technologies

Thermochemical conversion technologies use thermal energy in the form of heat to convert biomass into solid, liquid and gaseous products. The basic thermochemical conversion technologies are combustion, pyrolysis and gasification are principally classified based on the reaction temperature, availability of oxygen, residence time and type of product obtained.

6.1. Introduction to Combustion of Fuels

Fuel may be defined as a combustible substance available in bulk, which on burning in presence of air / oxygen generates heat that can be economically utilized for domestic and industrial purposes. The common fuels are compounds of carbon and hydrogen; in addition, variable percentages of oxygen and small percentages of sulphur and nitrogen are also present.

Biomass fuels are normally thermally degradable solids. Fuels may be directly delivered to the combustion furnace, but they are not combusted directly. These materials in a hot furnace are converted into volatile materials and carbonaceous chat, which burn with entirely different combustion characteristics. After mixing with air, the volatile products react with oxygen and burn with flames at the gaseous phase within the furnace space, whereas the carbonaceous char burns with glowing ignition at the solid phase on the grate. These two modes of combustion have entirely different chemical mechanisms and kinetics, which are dependent on both the chemical and physical compositions of the substrate and design of the combustion system.

Combustion of organic materials not only generates natural components of air such as carbon dioxide and water but also produces carbonaceous residues, smoke and tar and obnoxious gases of carbonyl derivatives, unsaturated compounds and carbon monoxide.

The various types of fuels like solid, liquid and gaseous fuels are available for firing in boilers, furnaces and other combustion equipments. The selection of right type of fuel depends on various factors such as availability, storage, handling, pollution and landed cost of fuel. The knowledge on the fuel properties helps in selecting the right fuel for the right purpose and efficient use of the fuel.

6.1.1. Direct combustion

Direct combustion is a mature and well established technology with numerous operating plants around the world. In combustion, the waste fuel is burnt in excess air in a controlled manner to produce heat. Flue gases from efficient combustion are mainly carbon dioxide and water vapour, with small amounts of other air emissions, depending on the nature of the biomass fuel. The flue gases are cleaned using flue gas scrubbers, bag filters and electrostatic precipitators, and if required, further chemical processing to reduce emission of oxides of nitrogen (NO_x) and other pollutants. The combustion heat is used to raise steam in a boiler. The steam is expanded through a turbine connected to a generator, thereby producing electricity. The combustion process is explained in Fig.6.1.

If there is a requirement for heat adjacent to the power plant, the energy plant can be configured as a cogeneration plant to generate electricity and provide heat simultaneously to nearby industries or other applications, such as district heating. This is done extensively in the sugar industry, where bagasse (crushed sugar cane residue) is used as fuel for providing the energy needs of the sugar mill.

Biomass fuels may also be co-combusted with fossil fuels in the existing power stations and cement kilns. This usually entails feeding relatively small quantities of biomass into the plant with the conventional fuel, without disturbing the operation of the plant. Grate furnaces are a common type of bioenergy combustion technology for converting woody biomass and bagasse to energy.

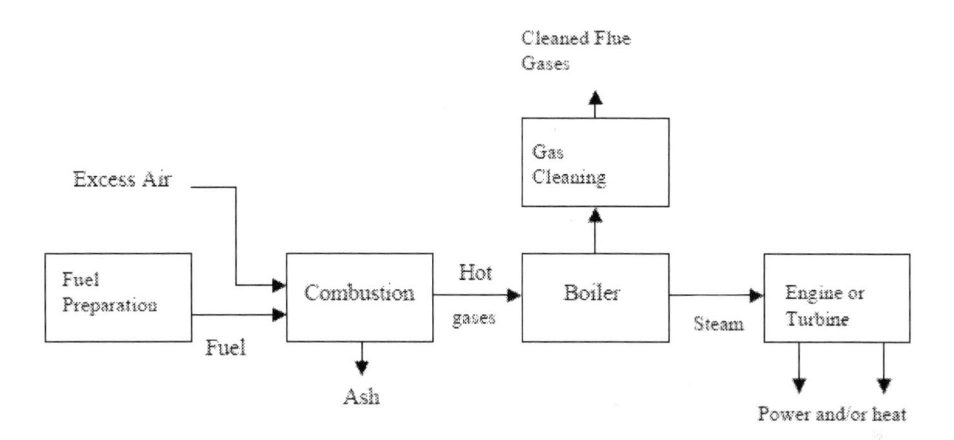

Fig.6.1. Direct Combustion System

6.2. Principles of combustion

Combustion refers to the rapid oxidation of fuel accompanied by the production of heat, or heat and light. Complete combustion of fuel is possible only in the presence of a sufficient supply of oxygen. Oxygen (O2) is one of the most common elements on Earth, making up 20.9% of our air. The rapid oxidation of the fuel results in a large amount of heat. Solid or liquid fuels must be converted to gas before burning. Heat is usually required to change liquids or solids into gases. Fuel gases will burn in a normal state if enough air is present.

Most of the 79% of air (which is not oxygen) is nitrogen with traces of other elements. Nitrogen is considered a temperature-reducing diluent that must be present to obtain the oxygen required for combustion.

Nitrogen reduces combustion efficiency by absorbing heat from burning fuels and diluting the flue gas. This reduces the heat available for transfer through the heat exchange surfaces. It also increases the volume of combustion by-products, which then have to travel faster through the heat exchanger and up the stack to allow additional fuel/air mixture to be introduced. This nitrogen can also combine with oxygen (especially at high flame temperatures) to form nitrogen oxides (NOx), which are toxic pollutants.

The carbon, hydrogen, and sulfur in the fuel combine with oxygen in the air to form carbon dioxide, water vapor, and sulfur dioxide, releasing 8,084 kcal, 28,922 kcal, and 2,224 kcal of heat, respectively. Under certain conditions, carbon can combine with oxygen to form carbon monoxide, which results in the release of a smaller amount of heat (2,430 kcal/kg carbon). Carbon burned to CO2 will produce more heat per unit of fuel than when CO or smoke is produced. Each kilogram of CO generated means a loss of 5654 kcal of heat (8084-2430).

$$C + O_2 \rightarrow CO_2 + 8084 \text{ kcal/kg of Carbon}$$

$$2C + O_2 \rightarrow 2\,CO + 2430 \text{ kcal/kg of Carbon}$$

$$2H_2 + O_2 \rightarrow 2H_2O + 28{,}922 \text{ kcal/kg of Hydrogen}$$

$$S + O_2 \rightarrow SO_2 + 2{,}224 \text{ kcal/kg of Sulphur}$$

6.2.1. 3 T's of combustion

The goal of a good burn is to release all the heat from the fuel. This is accomplished by controlling the "three T's" of combustion, which are (1) temperature high enough to ignite and sustain ignition of the fuel, (2) turbulence or complete mixing of the fuel and oxygen, and (3) time sufficient for complete ignition. combustion.

Commonly used fuels such as natural gas and propane are generally composed of carbon and hydrogen. Water vapor is a byproduct of hydrogen combustion. This removes heat from the flue gas that would otherwise be available for more heat transfer.

Natural gas contains more hydrogen and less carbon per kg than heating oils and as such produces more water vapor. As a result, more heat will be removed through the exhaust gases when burning natural gas. Too much or too little fuel with available combustion air can potentially lead to unburned fuel and carbon monoxide formation. A very specific amount of O2 is needed for perfect combustion, and some additional (excess) air is needed to ensure complete combustion. However, too much excess air will lead to heat and efficiency losses.

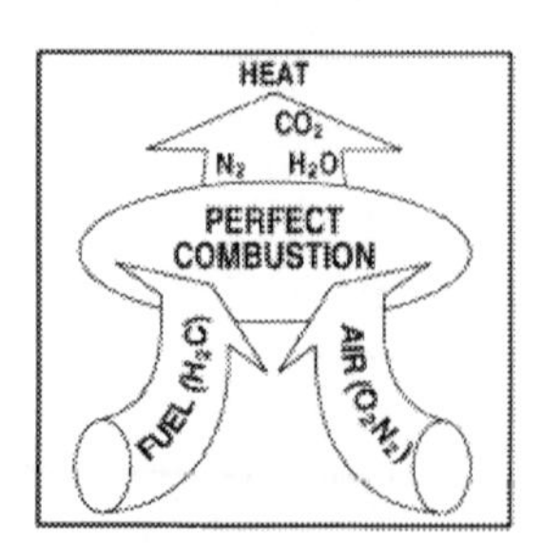

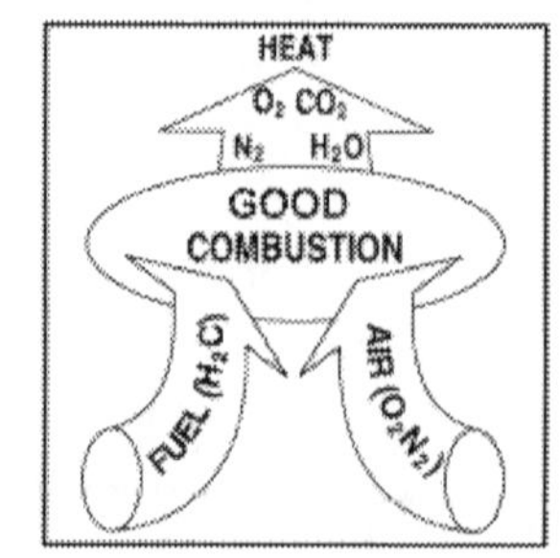

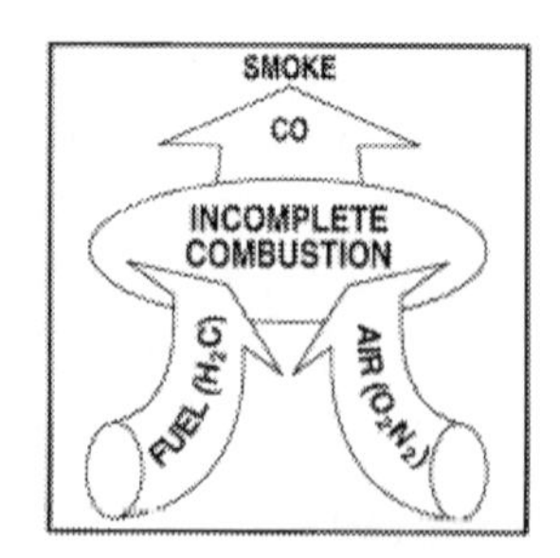

Fig.6.2. Types of combustion

Not all of the heat in the fuel is converted to heat and absorbed by the steam generating equipment. Typically, all the hydrogen in the fuel is burned, and most boiler fuels permitted under today's air pollution standards contain little or no sulfur. So the main challenge in combustion efficiency is focused on the unburnt carbon (in the ash or incompletely burned gas) that forms CO instead of CO2. The types of combustion are clearly shown in Fig. 6.2.

6.2.3. Combustion controls

Combustion control helps the burner to regulate fuel supply, air supply (fuel-air ratio) and flue gas exhaust to achieve optimum boiler efficiency. The amount of fuel fed to the burner must be proportional to the steam pressure and the amount of steam required. Combustion controls are also essential as a safety device to ensure safe operation of the boiler.

6.2.4. Conditions for efficient combustion

The following requirements are to be fulfilled for efficient combustion of any fuel in a furnace:

1. A sufficient amount of air (higher than stoichiometric amount of air) must be supplied in order to ensure the supply of sufficient oxygen for completion of the combustion reactions.

2. The fuel mass and the oxygen of the air must be in free and intimate contact.

3. While burning solid biomass and other solid fuels, the volatile combustion products leaving the fuel bed must be intimately mixed with the secondary air.

4. The volatile combustion products leaving the fuel bed must not be allowed to cool below the ignition point until the reactions are complete.

5. The volume of the furnace is to be designed in such a way that the provision for expansion of the gases during the process of combustion at high temperature is made.

6.2.5. Stochiometric calculation of air requirement

It is possible to predict the amount of air needed to burn one kg of fuel completely based on combustion equations. Equation for theoretical air requirement can be given by the equation,

$$\frac{\text{Kg of Air}}{\text{Kg of Fuel}} = 11.53C + 34.34(H_2 - O_2/8) + 4.29S$$

Where,

C, H_2, O_2 and S are the fractions, by weight, of each chemical constituent of the fuel.

6.2.5.1 Excess air

During combustion, insufficient air reduces fuel efficiency, creates highly toxic carbon monoxide gas, and produces soot. To ensure sufficient oxygen for complete reaction with the fuel, additional combustion air is usually supplied. This extra air, called "excess air", is expressed as a percentage of air above the amount theoretically needed for complete combustion. In actual combustion, the excess air required for gaseous fuels is typically about 15%.

$$\%\ \text{Excess Air} = \frac{\%\ O_2\ \text{measured}}{20.9 - \%O_2\ \text{measured}} \times 100$$

6.2.5.2. Combustion efficiency

Combustion efficiency is a measure of how efficiently the energy from the fuel is converted into useful energy. Combustion efficiency is determined by subtracting the heat content of the exhaust gases, expressed as a percentage of the calorific value of the fuel, from the total potential of the fuel and heat, or 100%.

$$\%\ \text{Combustion Efficiency} = 100\% - \left(\frac{\text{Stack heat losses}}{\text{Fuel heating value}} \times 100\right)$$

6.2. Types of Fuels

The fuels are classified into the following three types

1. Solid fuels

2. Liquid fuels

3. Gaseous fuels

6.3.1. Solid fuels

The different types of solid fuels are wood, peat, lignite, bituminous and anthracite coal.

6.3.1.1. Coal

Coal is a flammable black or brownish-black sedimentary rock used as a solid fossil fuel. It is primarily made up of 65 to 95% carbon and also contains hydrogen, sulphur, oxygen and nitrogen. It occurs in rock strata in layers or veins called coal beds or coal seams. A fossil fuel is the largest source of energy for the generation of electricity. Coal consists of an organic mass with some quantities of inorganic substances like water & mineral matter. Organic mass is a mixture of complex organic compounds of C, H, O, N and S. It also contains some absorbed gases, namely methane; CO_2 and CO. Carbon content of coal supplies most of its heating value and calorific value also very important.

The different types of coal are as follows,

1. Peat
2. Lignite
3. Sub- bituminous
4. Bituminous
5. Semi- anthracite
6. Anthracite

a. Peat

It is the first stage in the progress of transformation of buried vegetation into coal. Peat appears light brown and fibrous at the surface of peat deposits. But with increase in depth the colour becomes darker and the vegetable structure disappears. Peat contains high about 90% of moisture and small percentage of carbon and volatile matter.

The heating value of peat is less than 4000 kcal/kg, which is due to the presence of higher moisture content. Peat is not suitable for power plants. Peat is sun dried to remove the greater part of moisture and can be carbonized to produce gas and coke. Peat can be briquetted and used as a domestic fuel.

b. Lignite

Lignite is the second stage in the development of coal. Ranks lowest and is the youngest of the coals. On exposure to the atmosphere, the brown colour of lignite darkens and the moisture content reduces to an equilibrium value of 12 – 20%. On drying, lignite shrinks and breaks up readily in an irregular manner and cannot be transported far from the mine. Ignite spontaneously as it absorbs oxygen readily and must not be stored in the open.

Lignite is found in India, U.S.A, Germany, Australia & Canada. Major deposit of true lignite in Tamilnadu is in Neyveli (200 million tonnes) occurs about 60 m below the ground level & is 15- 20 m thick. Calorific value of air dried Neyveli coal is 4000-5500 kcal/kg. It can be used for manufacturing of producer gas and for the generation of electric power.

c. Sub bituminous

Sub bituminous is black in colour with a dull, waxy lustre. It is a non-coking coal, which breaks into smaller pieces on exposure to air. Sub bituminous is a variety of mature lignite resembling true coal in colour and appearance. Sub bituminous is denser and harder than lignite. Most sub- bituminous coals appear banded like bituminous coal and it is difficult to transport. Sub- bituminous coals occur in India, U.S.A, Germany, Australia and Canada. Some of the tertiary coals in Assam, Kashmir and Rajasthan are of the sub-bituminous type. Calorific value is 6800-7600 kcal/kg. It is used for power generation & space heating.

d. Bituminous

Most popular form of coal used for all purposes. It is a soft, dense, black coal, often has bands of bright and dull material in it. It is denser and harder than lignite and sub-bituminous coal. It does not disintegrate into slacks on exposure to the atmosphere. About 90% of the coal in this country falls in the bituminous and sub-bituminous categories. The calorific value of bituminous is 7500-9000 kcal/kg. It is used for generating electricity, making coke, gasification, domestic heating, etc.

e. Semi- anthracite

Semi-anthracite is intermediate between bituminous coal and anthracites. Semi-anthracite is harder than the most mature bituminous coal. Semi-anthracite ignites more easily than anthracite to give a short flame changing from yellow to blue.

Semi – anthracites do not occur in India. The calorific value of semi-anthracite is 8500-8800 kcal/kg.

f. Anthracite

Anthracite is the last stage in the process of transformation of buried vegetation into coal. It is the most mature and hardest coal. Anthracite is hard, black and lustrous. Anthracite has sub- metallic lustre, sometimes even a graphitic appearance. Although there is banded structure this is not always obvious and also zero caking power. Anthracite burns only at high temperature and does not fuse or soften. Anthracite is low in sulphur and high in carbon. Anthracite is the highest rank of coal. The calorific value of anthracite is 9000 kcal/kg. It is used for the following purposes,

1. For re-carbonizing steel,
2. For making carbon electrodes, brushes, battery parts, resistors, carbon refractory and corrosion resisting structural materials.
3. Used for filter and paint pigment.
4. For blending with coking coal to check its swelling and improving the coke quality.

6.3.2. Liquid fuels

Liquid fuels are those flammable or energy-generating molecules that can be used to create mechanical energy. Most liquid fuels come from fossil fuels. There are several types, such as hydrogen fuel (for use in automobiles), which are also classified as liquid fuel. In general, all crude oils are made of hydrocarbon compounds.

a. Crude oil

Crude oil is simply unprocessed oil found deep below the earth's surface. It can be clear to black in color and can be found as a liquid or solid. Crude oil is pumped and stored in barrels for future refining. The refining process can include filtering, adding additives, and specialized separation techniques to create specific crude oils and petroleum products. Oil contaminants are water, salts (Mg Cl2, CaCl2, NaCl, etc.) up to 2000-5000 mg/liter and sediments. Extra water in crude requires extra heat for its distillation; increase its cost of transportation, forms emulsion which absorbs materials like resin. Salt in crude oil causes scaling, corrosion and reduces heat transfer coefficient during its processing. Sediments present in crude causes erosion and scaling.

6.3.2.1. Components of petroleum

The petroleum consists of 3 main hydrocarbon groups and are as follows

a. Paraffins

These consist of straight or branched carbon rings saturated with hydrogen atoms, the simplest of which is methane (CH_4) the main ingredient of natural gas. Others in this group include ethane (C_2H_6), and propane (C_3H_8).

b. Naphthenes

Naphthenes consist of carbon rings, sometimes with side chains, saturated with hydrogen atoms. Naphthenes are chemically stable; they occur naturally in crude oil and have properties similar to paraffins.

c. Aromatics

Aromatic hydrocarbons are compounds that contain a ring of six carbon atoms with alternating double and single bonds and six attached hydrogen atoms. This type of structure is known as a benzene ring. They occur naturally in crude oil, and can also be created by the refining process.

6.3.2.2. Refining of petroleum

This process of condensation and vaporization takes place many times causing separation of the constituents of petroleum according to their boiling points. Thus, the higher boiling point fractions condense towards the lower part of the column and the lower boiling point fractions towards the upper part (Fig.6.3).

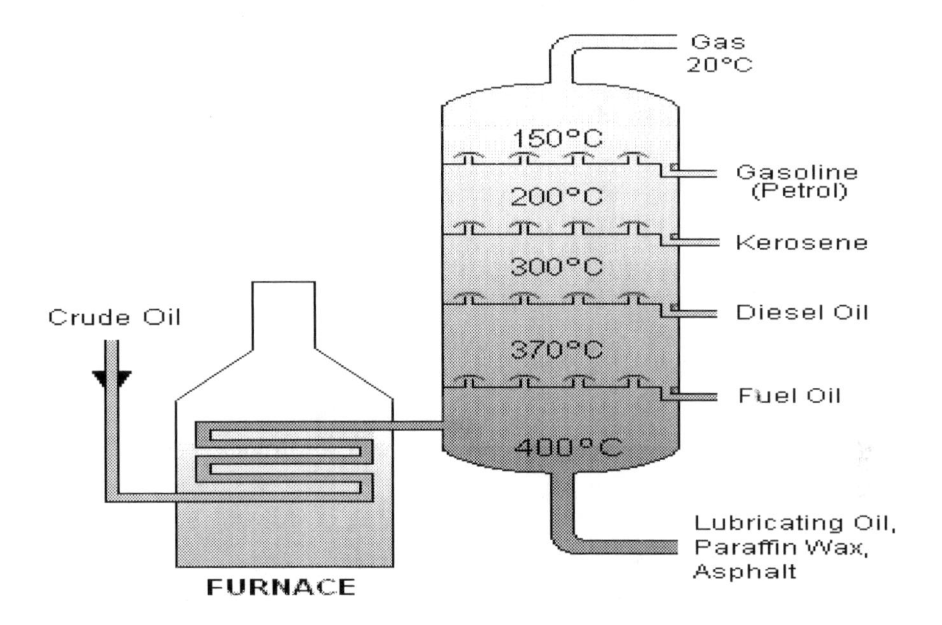

Fig.6.3. Refining of petroleum products through distillation process

The products of distillation are as follows:

1. Refinery gas (LPG) mainly consists of propane and butane (-160 to -40 °C)
2. Gasoline or petrol (30 -200 °C)
3. Naptha (120-200 °C)
4. Solvent Spirit (125-250 °C) or, Jet fuel (130-260 °C)
5. Kerosene (140-300 °C)
6. Gas oil (180-350 °C)
7. Lubricating oil (200-350 °C)
8. Petrolatum (220-350 °C)
9. Light fuel oil (> 200 °C)
10. Heavy fuel oil (> 250 °C)
11. Road making bitumen or tar
12. Wax and Residue pitch / coke.

6.3.3. Gaseous fuels

Gaseous fuels are those which are burnt in gaseous state in the air or oxygen to give heat for utilization in domestic / commercial sector. A remarkable feature of gaseous fuels is the absence of mineral impurities, consistency in quality and convenience in use. Gaseous fuels are termed as rich gas (if calorific value is > 4000 kcal/Nm^3) and lean gas (if calorific value is < 1500 kcal/Nm^3).

6.3.3.1. Classification of gaseous fuels

(a) Fuels naturally found in nature

» Natural gas,

» Methane from coal mines

(b) Fuel gases made from solid fuel

» Gases derived from coal

» Gases derived from waste and biomass

» Gases derived from other industrial processes (Blast furnace gas)

(c) Gases made from petroleum

» Liquefied Petroleum Gas (LPG)

» Refinery gases

» Gases from oil gasification

a. Natural gas

Natural gas is a fuel gas obtained from crude petroleum oil. Its boiling point ranges from -160 to -45°C. It is lighter and gets collected in the cavities of cap rocks and crude oil remains at the bottom. The natural gas composition is of Methane: 60 – 80%, Ethane: 5-8%, Propane: 18%, Heavier hydrocarbons (Propane, Butane, Pentane) 2-4%, and other non hydrocarbons such as N_2, H_2S, CO_2, etc. The calorific value of natural gas is 38 MJ/m^3.

Natural gas is considered as fuel of 21st century because of its greater efficiency and cost effectiveness. The world's largest proven gas reserves are located in Russia.

India presently produces natural gas of 87 million standard cubic meters per day. The main producers of natural gas in India are Oil & Natural Gas Corporation Ltd. (ONGC) and Oil India Limited (OIL).

b. Producer gas

The producer gas is a fuel obtained from low grade coals, peat, wood waste, coke, etc. The producer gas is obtained from chemical reaction between solid coal, air or and steam at temperature around 500°C. The producer gas from coke has a composition of 29% CO, 11% H_2, 55% N2, 0.5% CH_4, and 4.5% others. It is used in the production of iron, steel, coal gas, glass, refractory's, ceramics etc.

c. Liquefied petroleum gas (LPG)

Propane, butane, pentane are removed through refining from natural gas to form LPG. It is also called bottled gas. The LPG has a higher calorific value of 94 MJ/m^3. The uses of LPG include rural heating, motor fuel, refrigeration and cooking.

d. Refinery gas

The refinery gas is the gas produced in petroleum refineries by cracking, reforming, and other processes; principally methane, ethane, ethylene, butanes and butylenes.

e. Water gas

Water gas is a synthesis gas, containing carbon monoxide and hydrogen. It is a useful product but requires careful handling because of the risk of carbon monoxide poisoning. The gas is made by passing steam over a red-hot hydrocarbon fuel such as coke. Water gas had a low calorific value than coal gas so the calorific value was often boosted by passing the gas through a heated retort into which oil was sprayed. The boosted gas is called carbureted water gas. The carbureted water gas has a calorific value of 19 MJ/m^3.

f. Blast furnace gas

Blast furnace gas is a low grade, dusty by-product of blast furnaces that is generated when the iron ore is reduced with coke to metallic iron. Blast furnace gas has a very low heating value about 3.48 MJ/m^3 at NTP. The composition of blast furnace gas is of 60% N_2, 27% CO, 11% CO_2 and 2% H_2. Blast furnace gas is used for heating stoves of blast furnace, to pre-heat the blast, reheating furnaces, pre-heating water, heating coke fed to furnace.

g. Coke oven gas

This gas produced in coke ovens during high temperature carburization of coal (1000 to 1200°C). The coke oven gas is produced intermittently in batches at interval of 10 hours. The composition of coke oven gas is of 48% H_2, 32% CH_4, 8% N_2, 6% CO: 6% and other trace gases.

6.4. Improved Chulhas

In India, 80% of total population lives in villages. The major energy demand of rural population is for cooking which contributes to about 98% of their total energy consumption. Wood, agricultural waste and biomass are used as fuel in rural kitchen. The cooking appliances which are commonly used in rural houses have very low thermal efficiency (6-8%) and hence per capita energy consumption in rural areas is much higher than that in urban areas.

Conventional stoves waste a lot of energy and pose many pollution hazards. Most traditional stoves can utilize only 2 – 10% of the energy generated by the fuel. The growing gap between availability and demand for fire wood, poor thermal performance and pollution caused by traditional stoves forced the technologists to concentrate their attention on improving the thermal efficiencies of stoves.

6.4.1. Single pot chulha

The single pot chulha has a double wall with a gap of 2.5 cm. It has a grate at the bottom of the combustion chamber (Fig.6.4). Legs have been provided in the four corners of the chulha (5 cm height) as the ash can be collected below the grate. The outer wall has two rectangular secondary air openings at the lower portion on both sides for air entry. The inner wall has 1 cm diameter holes which maintain a triangular pitch of approximately 3 cm. The secondary air enters through the rectangular opening in the outer wall, gets heated in the annular chamber and while moving up it passes through the holes in the combustion chamber. The preheated air helps in proper burning of the fuel. The efficiency of single pot improved chulha is 24%.

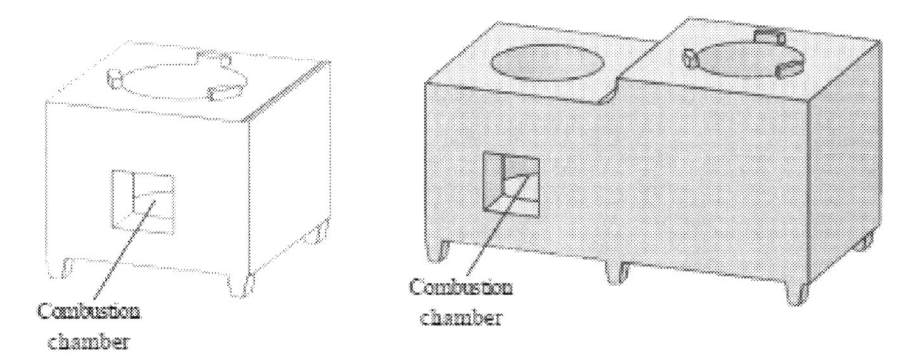

Fig.6.4. and 6.5. Single and Double pot chulha

6.4.2. Double pot improved chulha

The TNAU double pot portable chulha (chimneyless) is made with two walls (Fig.6.5). Around the first pot, an annular chamber having a width of 2.5 cm is left and the outer wall is constructed. The outer wall is also extended to cover the second pot in which case the annular chamber width is 3.5 cm, because of the smaller diameter of the second pot hole. Two secondary air inlets are made, one on the outer wall with rectangular shape (17 cm x 1 cm) near the combustion chamber and the other at the bottom of the second pot hole with round shape having a diameter of 5 cm. At the bottom of the first pot hole in the base, a hole of diameter 14 cm is made and a grate (C.I.) is placed over it. For the entry of secondary air to the first pothole, 1 cm dia holes are made with a triangular pitch of 3 cm on the inner side of first pot hole and also on the tunnel projecting into the second pot hole. The efficiency of single pot improved chulha is 26%.

6.4.3. TNAU Noon Meal Improved Chulha (Community Chulha)

This community model (TNAU midday meal chulha) double pot chula is made from local materials like clay, sand, paddy husks etc. This chulha has two air holes (6 x 9 cm) on the sides of the hearth mouth for better combustion (Fig.6.6). Tunnels were provided to pass the flame to the second pot and carry smoke easily into the chimney pipe. The height of the first pot hole is made lower (40 cm) than the second pot hole (48 cm) for the better propagation of flame.

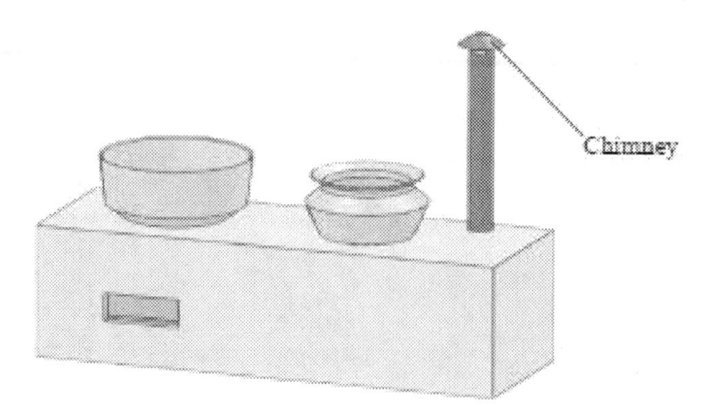

Fig.6.6. TNAU Noon meal improved chulha

6.4.4. Biomass gas stove

The biomass gas stove has been developed for small scale thermal application in Agriculture and allied industries. This stove widens the market for agro wastes, makes possible a higher efficiency and in some cases reduces the time and investment, all by comparison with combustion. The biomass gas stove is a natural convection type updraft gasifier consisting of a cylindrical body made of clay, sand and paddy husk with its top open and bottom closed. The diameters and height of the stove are 290mm and 630mm respectively (Fig.6.7). This can be reduced depending on the applications. An iron grate to hold the biomass is fixed at 50 mm from the base of the reactor. The bottom is provided with an air opening cum ash removal door. At the top, provision is made to place vessel for cooking, boiling etc. The biomass viz., wood chips, agricultural residues like coconut shell, groundnut shell, arecanut husk, tree barks and leaves can be used in this biomass gas stove. The feedstock used should preferably bc in the form of small chips, splinters and small logs.

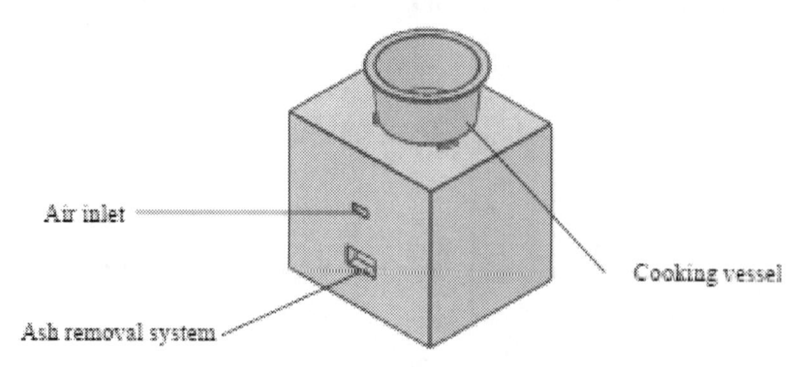

Fig. 6.7. Biomass gas stove

The stove can be ignited in two methods. One is igniting from bottom and other is from the top. In the first method, a bed of ignited charcoal (½- 1 kg) to a depth of 50mm is placed over the grate. Then the biomass used is dumped over the red hot charcoal. A white gas emanates first and the following the combustible gas is ignited. Whereas in the second method, it is ignited at the top with diesel soaked waste cotton or by using easily combustible agricultural residues. Initially burning is started in combustion mode and then changes to gasification mode. The stove consumes 4 to 5 kg of biomass per hour. The quantity varies with the type of biomass used. The efficiency of single pot improved chulha is 25%.

6.5. Pyrolysis

Pyrolysis is defined as the thermal degradation of biomass at the temperature of about 250 – 900 °C in the absence of oxygen to yield solid, liquid and gaseous products. The solid product so obtained is called as charcoal / biochar which is rich in carbon content up to 60 – 80% and thus has more calorific value. Oil from condensable vapors at room temperatures is called pyrolysis oil (also known as bio-oil, bio-crude) and non-condensable gases such as carbon monoxide, carbon dioxide, hydrogen and low molecular weight hydrocarbon gases such as methane, collectively called synthesis gas (syngas or generator gas).

6.5.1. Types of pyrolysis process

Depending upon the temperature and time of the pyrolysis process, the product yield percentage is given below in table 6.1.

1. Slow Pyrolysis (250 – 400 °C)
2. Intermediate Pyrolysis (350 – 450 °C)
3. Fast Pyrolysis (450 – 900 °C)

Table 6.1. Product yield (%) during pyrolysis process

S.No	Products	Slow Pyrolysis	Intermediate Pyrolysis	Fast Pyrolysis
1.	Biochar (%)	35	25	12
2.	Bio-oil (%)	30	50	75
3.	Producer gas (%)	35	35	13

In general, higher percentage of solid product (char) will be produced from slow pyrolysis and higher percentage of liquid product (bio-oil) will be produced from intermediate and fast pyrolysis.

6.5.2. Products of pyrolysis process

When the biomass is thermally decomposed between 250 and 900 °C in absence of oxygen, three kinds of products namely biochar or charcoal, bio-oil and producer gas will be obtained. The production process and applications of the products are described in detailed manner.

6.5.2.1. Biochar

The solid products include char which forms on surface of reacting particles and the residue after the pyrolysis. Depending on temperature, the char fraction contains inorganic materials ashed to varying degrees, any unconverted organic solids and carbonaceous residues, produced on thermal decomposition of the organic components, particularly lignin.

a. Production of biochar

The biomass substances like agro and wood residues such as Prosopis, casuarina, coconut shell and saw dust etc., are thermally degraded in the absence of oxygen condition at temperatures between 250 – 400 °C to yield a solid product called biochar. The reactor used to produce biochar is called slow pyrolyzer. The biochar yield will vary between 30 -35%. Biochar retains ~ 20% of the weight and 30% of the energy of the biomass, and hence ~70% of the energy is released as usable vapours.

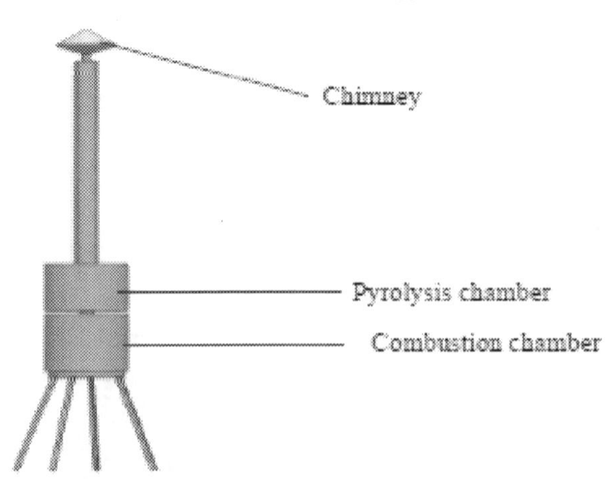

Fig.6.8. Biochar production unit

The slow pyrolyzer consists of bottom chamber, top chamber and chimney unit (Fig.6.8). The bottom chamber is called combustion unit, where charcoal is used as the fuelling source. The top chamber is covered with the biomass materials where the pyrolysis process takes place to produce biochar. The heat is passed from bottom chamber to top chamber and volatiles were released through the chimney set up at the top. Once the reactor is cooled to atmospheric temperature, the biochar is collected and used for several applications.

b. Characteristics of biochar

- High carbon content (60 – 90%)
- Heating value 25-30 MJ/kg can be used as a fuel.
- Resistant to biodegradation
- Significant adsorptive qualities
- Nutrients essentially lock on to the structure
- Increases water retention, moisture holding capacity and soil organic carbon
- Enhances microbial biomass and soil biology
- Reduces fertilizer requirements by improving retention of soluble nutrients
- Improves nutrient retention by increasing cation exchange capacity (CEC) by about 50% and reduces nutrient runoff to water ways
- Adjusts soil pH
- Reduces N leaching by 60%, increases crop productivity by 38 to 45%, saving 20% fertilizer and 10% savings in irrigation and seeds

c. Applications of biochar

1. It can be used as a soil conditioner, thus it mitigates green house gas emissions
2. It can be used as a substitute fuel
3. It can be as adsorbent to treat waste water in the waste water treatment plants

6.5.2.2. Bio oil

Bio oil is the liquid condensate of the vapors of a pyrolysis reaction. It is a dark brownish viscous liquid that bears some resemblance to fossil crude oil. It is at times marketed as "liquid smoke." However, bio-oil is a complex oxygenated compound comprised of water, water soluble compounds, such as acids, esters, etc., and water-insoluble compounds, usually called pyrolytic lignin because it comes from the lignin fraction of the biomass.

a. Production of bio oil

Pyrolysis of organic materials such as biomass at high temperatures (above 450°C) decomposes the fuel source into charcoal (carbon and ash) and volatiles. Pyrolysis itself does not normally release excessive heat; rather, it requires heat to sustain. This includes condensable vapors called pyrolysis oil (also known as bio-oil, bio-crude, etc.) at room temperatures and non-condensable (permanent) gases such as carbon monoxide, carbon dioxide, hydrogen and low molecular weight hydrocarbon gases such as methane, collectively called synthesis gas (syngas or generator gas). The characteristics of bio oil are listed in Table 6.2.

Table 6.2. Characteristics of bio oil

S.No	Properties	Percentage
1.	Moisture (%)	15 - 30
2.	pH	2.8 – 4.0
3.	Specific gravity	1.1 – 1.2
Composition of Bio-oil		
4.	Carbon (%)	55 – 64
5.	Hydrogen (%)	5 – 8
	Nitrogen (%)	0.05 – 1.0
	Ash (%)	0.03 – 0.30
	Heating Value (MJ/Kg)	16 – 26
	Viscosity, cP 104 °F and 25% water	25 - 100

b. Applications of Bio-oil

1. Bio-oil can be a substitute for fossil fuels to generate heat, power and/or chemicals.

2. Boilers and furnaces (including power stations) can be fueled with bio-oil in

the short term; whereas, turbines and diesel engines may become available on the somewhat longer term. Upgrading of the bio-oil to a transportation fuel is technically feasible but needs further development.

6.5.3. Types of pyrolysis reactors

Pyrolysis can be classified as slow, medium or fast pyrolysis. Slow pyrolysis is an ancient art whose origins can be traced back to the time when man created fire. It maximizes charcoal yield and is used to produce charcoal for use as fuel in developing countries. Although pyrolysis does not release heat, the slow pyrolytic reaction to convert biomass mainly into charcoal can be slightly exothermic (releasing heat), so large quantities of charcoal can be produced by burying the biomass underground and starting the reaction.

Currently, most interest in pyrolysis focuses on fast pyrolysis, which maximizes liquid rather than charcoal production. It involves fast heating rates and higher temperatures than slow pyrolysis and is endothermic. The ideal conditions for fast pyrolysis involve rapid heating rates (> 240° C per second) and temperatures usually in excess of 537° C.

Due to its short residence time, the main products from biomass fast pyrolysis are ethylene-rich gases that can be condensed to produce bio-oil and alcohols rather than char and tar. Char or charcoal is produced from biomass that can store carbon. Interest in it is growing due to concerns about global warming caused by emissions of CO2 and other greenhouse gases. The different types of pyrolysis reactors for producing bio-oil are circulating fluidized bed, bubbling type fluidized bed, rotating cone and rotary pyrolysis reactors.

6.5.3.1. Bubbling fluidized bed pyrolyzer

In a fluidized bed pyrolyzer, a heated sand medium in a zero-oxygen environment rapidly heats the feedstock (biomass) to 400-900°C, where it decomposes into solid semi-coke, gas, vapors and aerosols that exit the reactor by transport fluidization. gas stream (Fig. 6.9). After leaving the reactor zone, the charcoal can be removed by a cyclone separator and stored. Washed gases, vapors and aerosols enter the direct extinguishing system, where they are rapidly cooled (< 51 °C) directly with liquid immiscible with bio-oil or indirectly using coolers (heat exchanger). The condensed bio-oil is collected and stored, and the non-condensable gas (syngas) can be recycled or used as fuel for heating the reactor. High liquid yields (about 60% weight of biomass on a dry basis) can be typically recovered.

6.5.3.2. Circulating fluidized bed pyrolyzer

These reactors (Fig.6.10) are similar to bubbling fluidized bed reactors but have shorter residence times for chars and vapors. Short residence times in the reactor result in higher gas velocities, faster escape of steam and semi-coke, and higher semi-coke content in the bio-oil than fluid bed bubbling. However, they have higher processing capacity, better gas-solid contact, and improved ability to handle solids that are more difficult to fluidize, but are less commonly used. The heat input typically comes from the secondary combustion chamber.

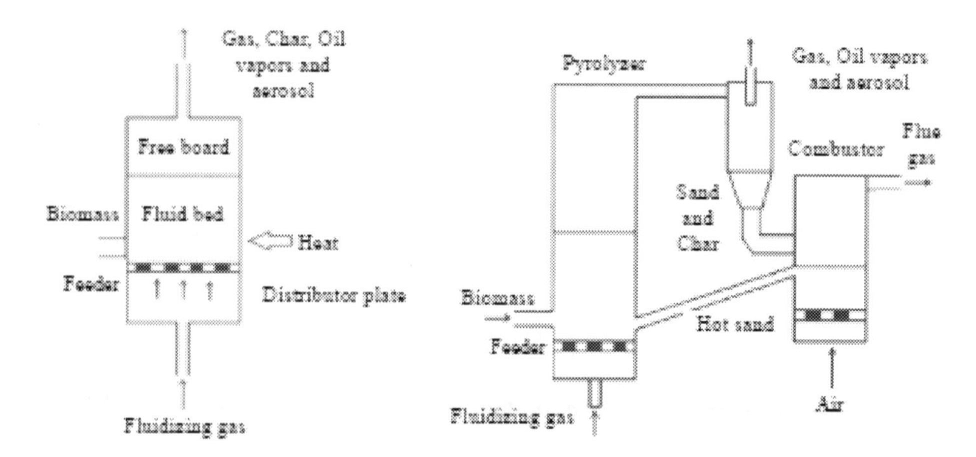

Fig.6.9. Bubbling fluidized bed reactor **Fig.6.10. Circulating fluidized bed reactor**

6.5.3.3. Rotating plate pyrolysis reactor

Rotating plate pyrolysis reactors (Fig. 6.11) operate while under pressure, the heat transferred from the hot surface can soften and vaporize the feedstock in contact with it, allowing the pyrolysis reaction to move the biomass in one direction. With this arrangement, larger particles, including logs, can be pyrolyzed without pulverizing them. The most important feature is that there is no requirement for an inert gas medium, thereby resulting in smaller processing equipment and more intense reactions. However, the process is dependent on surface area, so scaling can be an issue for the larger facilities.

6.5.3.4. Rotating cone pyrolysis reactor

During the pyrolysis process, biomass particles at room temperature and hot sand are introduced near the bottom of a cone at the same time. They are mixed and transported upwards by the rotation of the cone. The pressures of outgoing materials are slightly above atmospheric levels. Rapid heating and short gas phase residence times can be easily achieved in this reactor (Fig.6.12).

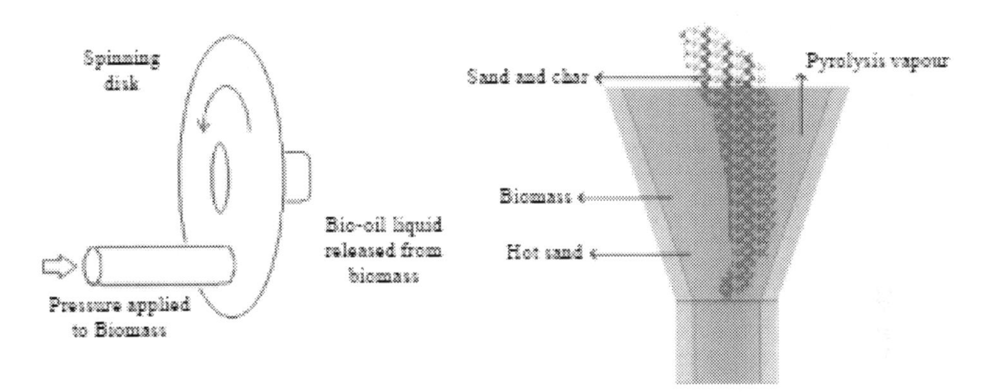

Fig.6.11. Rotating plate pyrolysis **Fig.6.12. Rotary plate pyrolysis**

6.6. Benefits of the Pyrolysis Process and its Products

Pyrolysis of agricultural residues can help meet renewable energy goals by displacing fossil fuels and thereby addressing global warming concerns.

i. Pyrolysis offers more scope for the use of products from agricultural waste than its incineration.

ii. When agricultural residues are first pyrolyzed, bio-oil, gases and bio-char can not only be used as fuel, but can also be purified and used as feedstock for petrochemical and other applications. The use of biochar for soil amendment and as a carbon sequestration and climate mitigation agent is gaining worldwide attention.

iii. Transportation fuels such as methanol and Fischer-Tropsch liquids (synthetic diesel) can be obtained from bio-oil by using bio-oil as gasification feedstock instead of bulky biomass and can save some transportation costs. In addition, there is a wide variety of chemicals that can be extracted or obtained from bio-oil.

6.7. Comparison between Slow, Intermediate and Fast Pyrolysis

The Table 6.3. represents the comparison between slow, intermediate and fast pyrolysis in terms of product yield characteristics.

Table 6.3. Comparison between typical pyrolysis systems

Technology	Temperature (°C)	Residence Time	Char (%)	Liquid (%)	Gas (%)
Slow pyrolysis	~ 400 °C	h-weeks	35	30	35
Intermediate pyrolysis	~500 °C	~10-20 secs	20	50	30
Fast pyrolysis	~500 °C	~1sec	12	75	13

6.8. Gasification

Gasification is a thermo-chemical process in which, biomass is thermally degraded with restricted amount of air/ oxygen to yield a mixture of combustible gases consisting of carbon monoxide, hydrogen and traces of methane called producer gas. The reactor where the gasification takes place is called as gasifier.

6.8.1. Importance of gasification

Generally, the biomass consists of atoms of carbon, hydrogen and oxygen. When complete combustion takes place, carbon present in the biomass to produce carbon dioxide, hydrogen is converted into water vapour. But, when partial oxidation takes place, the biomass is directly converted into mixture of gas such as carbon dioxide, hydrogen, nitrogen, carbon monoxide and methane is called as producer gas. Hydrogen, methane and carbon monoxide are the combustible gases. This producer gas can be directly used in the generator to produce electricity. Otherwise, wood materials can be used in the gasifier to produce methanol which has both engine and chemical applications.

6.8.2. Principle of gasification process

Biomass gasifier is simple in design and vertically cylindrical in shape. It consists of grate, air supply system with blower and gas scrubbing unit. Based on the temperature, the reactor consists of four zones (Fig. 6.13). They are,

1. Drying zone (100 – 150 °C)
2. Pyrolysis zone (150 – 700 °C)
3. Oxidation zone (700 – 1400 °C)
4. Reduction zone (800 – 1100 °C)

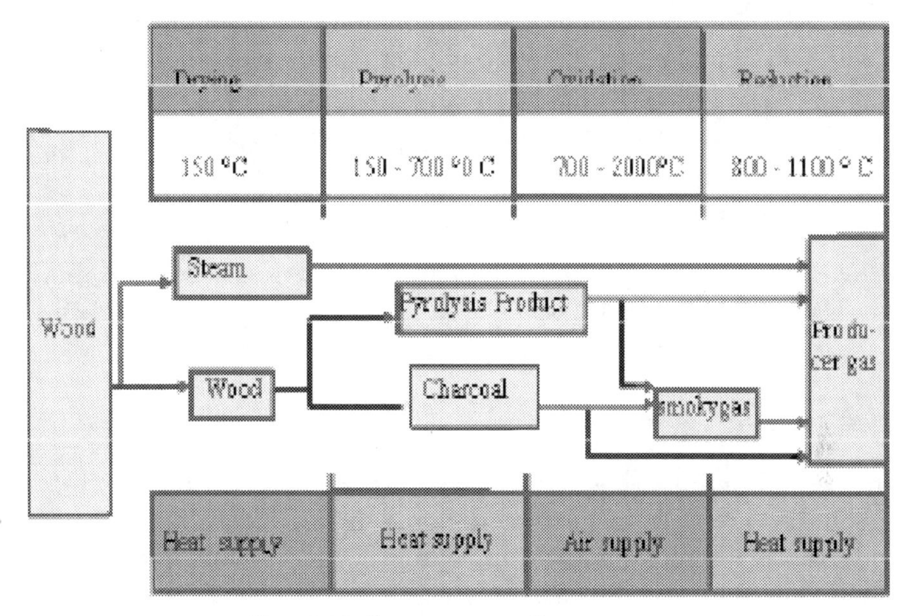

Fig. 6.13. Stages in gasification

6.8.2.1. Drying zone

In this zone, biomass are fed with the moisture content ranged from 10-15%. At temperatures between 100 and 150 °C, the water present in the biomass is converted into steam.

6.8.2.2. Pyrolysis zone

It is the process wherein heat is used to breakdown biomass in the absence of air to yield charcoal, wood-oils, tars and gases. In this zone, the solid material starts disintegrating at 150 - 700°C to produce char as well as condensable and non-condensable gases.

6.8.2.3. Oxidation zone

In this zone, when carbon in the biomass is reacted with air / oxygen which is converted into carbon dioxide, hydrogen with oxygen to form steam. Apart from heat generation, all condensable and organic products of pyrolysis get converted and oxidized. The process in this zone is exothermic and temperature will be about 1400°**C.**

6.8.2.4. Reduction zone

In this zone, carbon dioxide and steam is reacted with oxygen and reduced to carbon monoxide and hydrogen. In this zone, sensible heat of gases and charcoal is absorbed in endothermic reactions. The temperature at this zone is about 800 – 1100°**C.**

6.9. Types of gasifiers

Based on the bed, gasifier is classified into fixed bed gasifier and fluidized bed gasifier. And based on the flow and direction of air or fuel, fixed bed gasifier is classified into updraft gasifier, downdraft gasifier and cross draft gasifier.

6.9.1. Updraft gasifier

The up draft gasifier (Fig.6.14) is characterized by a counter current flow of fuel and air or gas in the reactor. The producer gas exits from the top of the gasifier. The biomass enters the top of the reaction chamber while steam and air or oxygen enters from bottom through grate. The fuel flows downward and up-flowing hot gases pyrolyze it. Some of the resulting charcoal residue falls to the grate, where it burns, producing heat and giving off carbon dioxide and water vapor. The CO_2 and H_2O react with other charcoal particles, producing carbon monoxide and hydrogen gases. The gases exit from the top and ashes fall through the grate. The updraft design is relatively simple and can handle biomass fuels with high ash and moisture content. The gas contains 10-20% volatile oils making the gas unsuitable for use in engines or gas turbines. These gasifiers are best suited for thermal applications.

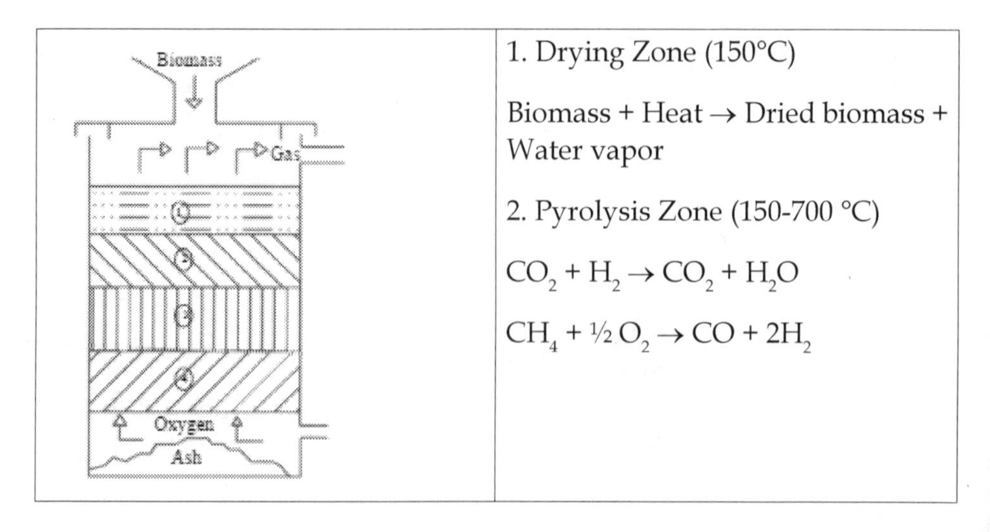

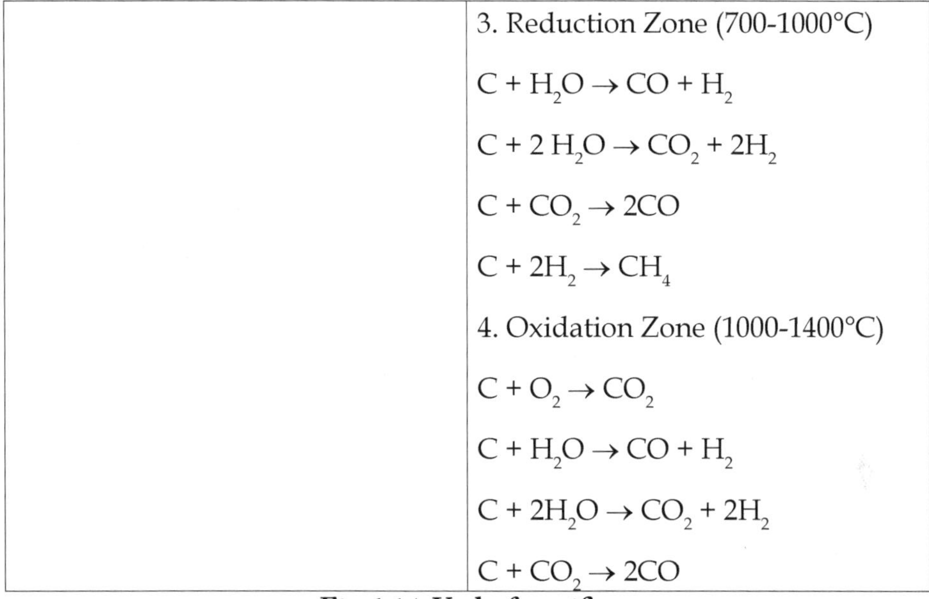

Fig.6.14. Updraft gasifier

6.9.2. Downdraft gasifier

The down draft gasifier (Fig.6.15) is characterized by co-current flow of air or gas and the fuel. Fuel and air or gas move in the same direction. Gasifier requires drying of biomass fuel to a moisture content of less than 20%. Fuel and air or oxygen enters the top of the reaction chamber. Down-flowing fuel particles ignite, burning intensely and leaving a charcoal residue. The charcoal which is about 5 to 15% of the mass of original fuel then reacts with the combustion gases, producing CO and H_2 gases. These gases flow down and exit from the chamber below a grate. The producer gas leaving the gasifier is at a high temperature. Combustion ash falls through the grate. The advantage of the downdraft design is the very low tar content of the producer gas and it can be used for power production.

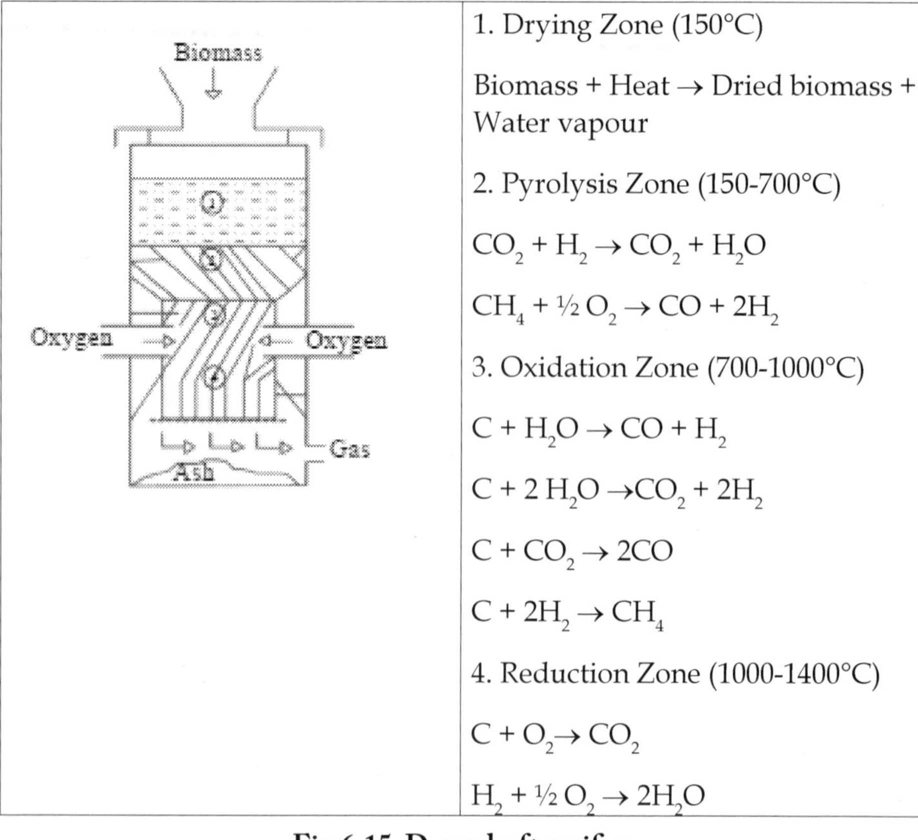

1. Drying Zone (150°C)

Biomass + Heat → Dried biomass + Water vapour

2. Pyrolysis Zone (150-700°C)

$CO_2 + H_2 \rightarrow CO_2 + H_2O$

$CH_4 + \frac{1}{2} O_2 \rightarrow CO + 2H_2$

3. Oxidation Zone (700-1000°C)

$C + H_2O \rightarrow CO + H_2$

$C + 2 H_2O \rightarrow CO_2 + 2H_2$

$C + CO_2 \rightarrow 2CO$

$C + 2H_2 \rightarrow CH_4$

4. Reduction Zone (1000-1400°C)

$C + O_2 \rightarrow CO_2$

$H_2 + \frac{1}{2} O_2 \rightarrow 2H_2O$

Fig.6.15. Downdraft gasifier

6.9.3. Cross draft gasifier

This gasifier can operate with wide variety of fuels compared to an up draft or a down draft gasifier. It has higher gas velocity at the gas exit with higher temperature, poor CO_2 reduction are certain characteristics of this type of gasifier. This type of gasifier has been used for gasification of coal.

6.9.4. Fluidized bed gasifier

Fluidized bed gasifier has a homogeneous reactor bed of some inert sand material. The fuel is introduced in the inert bed material and air passed at the bottom of the bed in the reactor. This gasifier is characterized by high gas exit temperature, very high solid particulate matter in the gas and relatively low efficiency. The gasifier can operate with low bulk density materials such as agro-residues, leaves, etc. A fluidization bed is a chamber with a perforated floor having pressurized air flowing vertically where a particle medium usually sand, is contained. The pressurized and flowing air helps the medium allowing it to act as a fluid.

This type of reactor can be used for complete combustion applications which would allow it to use liquid wastes such as used engine oil, non-recyclable plastics & old shoes, garbage for generation of heat. It has greater potential to treat municipal waste such as garbage. It is quicker in response with a shorter start time. The design of the system is very complex design in nature.

6.9.4. Characteristics of producer gas

The composition of the producer gas is purely depending on the type of biomass used and type of gasifiers. The producer gas consists of 22% carbon monoxide, 12% hydrogen, 2% methane, 9% carbon monoxide and 55% nitrogen. The heating value of 1 m^3 producer gas is about 1100 – 1200 kcal/m^3. Generally, from 1 kg of biomass, 2.5 m^3 of producer gas will be obtained. The factors such as moisture content, bulk density, ash content, tar content, air supply affects the performance of the gasifier.

6.10. Factors Influencing the Performance of the Gasifier

There are many factors which are affecting the performance of the gasification process. They are listed below.

- » Energy content
- » Size and shape of the fuel
- » Bulk weight or calorie value per volume
- » Tar content
- » Moisture content
- » Dust tendency
- » Ash and slag tendency
- » Reaction response
- » Equivalence ratio
- » Reactor pressure
- » Temperature

6.10.1. Equivalence ratio

The equivalence ratio, ϕ is defined as the ratio of the actual air supplied to the stoichiometric air required. Equivalence ratio for gasification process is 0.2 - 0.4. It can be expressed as follows:

$$\phi = \frac{\text{Actual air supplied for complete combustion}}{\text{Stoichiometric air required for complete combustion}}$$

6.10.2. Turn down ratio

Turn down ratio is defined as the ratio of maximum rated output to minimum rated output. It was calculated using the following formula.

$$\text{Turn down ratio} = \frac{\text{Maximum rated output, m}^3\ \text{h}^{-1}}{\text{Minimum rated output, m}^3\ \text{h}^{-1}}$$

6.10.3. Specific gasification rate

The specific gasification rate is the rate of fuel consumption per unit reactor area. It was calculated using the following formula

$$\text{Specific gasification rate, kgm}^{-2}\text{h}^{-1} = \frac{\text{Feedstock consumption rate, kg h}^{-1}}{\text{Cross sectional area of reactor, m}^2} \times 100$$

6.10.4. Hot gas efficiency

The hot gas efficiency was calculated as the ratio of heat energy in producer gas and heat content in feedstock. The hot gas efficiency is given by

$$\eta_h = \frac{\text{Heat energy in gas}}{\text{Heat input by feed stock}} \times 100$$

$$\eta_h = \frac{\text{Rated gas flow, m}^3\ \text{h}^{-1} \times \text{CV of gas, MJ m}^3}{\text{Feedstock consumption rate, kg h}^{-1} \times \text{CV of feedstock, MJ kg}^{-1}} \times 100$$

6.10.5. Cold gas efficiency

The cold gas efficiency was calculated as the ratio of heat energy in producer gas and heat content in feedstock along with input energy consumed. The cold gas efficiency is given by

$$\eta_c = \frac{\text{Rated gas flow, m}^3\text{ h}^{-1} \times \text{CV of gas, MJ m}^3}{\text{Feedstock consumption rate, kg h}^{-1} \times \text{CV of feedstock, MJ kg}^{-1} + \text{Input energy}} \times 100$$

6.11. Applications of Gasifier

Gasifiers can be used to produce heat directly for process applications. It can be coupled to internal combustion engines either to produced shaft power or electrical energy using an alternator. Small-scale gasifiers find wide application in rural area for energizing diesel engine pumps. Medium scale gasifiers can be used in rural industries for thermal and shaft power applications. These gasifiers can also be used to power vehicles. Large-scale gasifiers find application in industries as a source of electrification and they are run with specific feed stocks like wood.

Gas produced from biomass can be used for production of heat and electricity. In these cases the gasifying agent is usually air. The gas can be used also as a new syngas for production of methanol, ammonia or substitute natural gas. This requires a gas free of N_2 and other inerts. Gasification with O_2 is the usual method, but other techniques are available too.

a. Direct heating

There is considerable scope and potential for the use of gas produced by gasification in heat exchangers and boilers in the process industries. Some of its uses are to fire open hearth furnaces, cement kilns, tunnel kilns for heavy clay industries, ceramic processes, glass melting, agricultural and industrial driers. After removal of ash and char particles in one or more cyclones, the producer gas can be fired in an oven, kiln or furnace for production of steam or hot water or for calcining or roasting operations. Retrofitting combustion equipment is possible and does not necessarily lead to derating. Potential losses in radiation sections are often compensated by an increased heat transfer in convective sections. The erection of a completely new 0.38 MW gasifier / boiler combination was calculated to be less costly than a direct combustion system. Finally, the producer gas can be used for drying purposes, for instance for drying of the wet biomass to be gasified.

A further advantage is that producer gas can directly replace natural gas, town gas or oil with simple modification of the boiler heat transfer areas and burner equipment (for cooking etc.). The use of gas produced by gasification as a heat source can be applied in both small and large scale systems provided power is available to draw gases through them.

b. Power generation

Besides the steam boiler, steam engine or turbine power can be generated by combustion of the gas in an internal combustion engine or gas turbine. For this purpose the gas must be cleaned thoroughly. For use in reciprocating engines the gas should contain less than 20 mg Nm^3 of dust and tar constituents. Particles, especially those with dimensions over 3 mm have to be eliminated. Approximately 0.8 kWh of net power is generated from one kilogram of air-dry biomass. It is possible to convert standard spark ignition and diesel engines; changes are simple and usually are limited to items such as the gas intake system, gas-air mixer, and spark plugs. The low heating value of the gas / air mixture results in 30-40% derating of engines normally fueled by gasoline; this is overcome by supercharging. A diesel engine shows a much smaller derating of 10% because of its high excess air ratio, but they always need a small quantity of diesel oil for the ignition of the gas / air mixture. This quantity is approximately 15% of its maximum energy input. If the gas cleaning system is well designed there will be no extra wear or maintenance of engine parts.

Stationary units for electricity generation may be economically viable up to 1 MWe, but much depends on site-specific factors. Producer gas systems for traction are used only in remote areas, where the time-consuming operation and maintenance are balanced by the high fuel prices.

Although gas turbines, preferably in combination with pressurized producer gas systems, seem to be very attractive for power generation, no breakthrough has taken place until today. Since it is preferred to fire the hot gas directly into the combustion chamber, an extremely effective hot gas cleaning is required. A turbine is not only sensitive to particles but also to attack of vapour phase components such as potassium oxide, sulphides, and chlorides. The schematic diagram of the gasifier power plant is given in Fig.6.16.

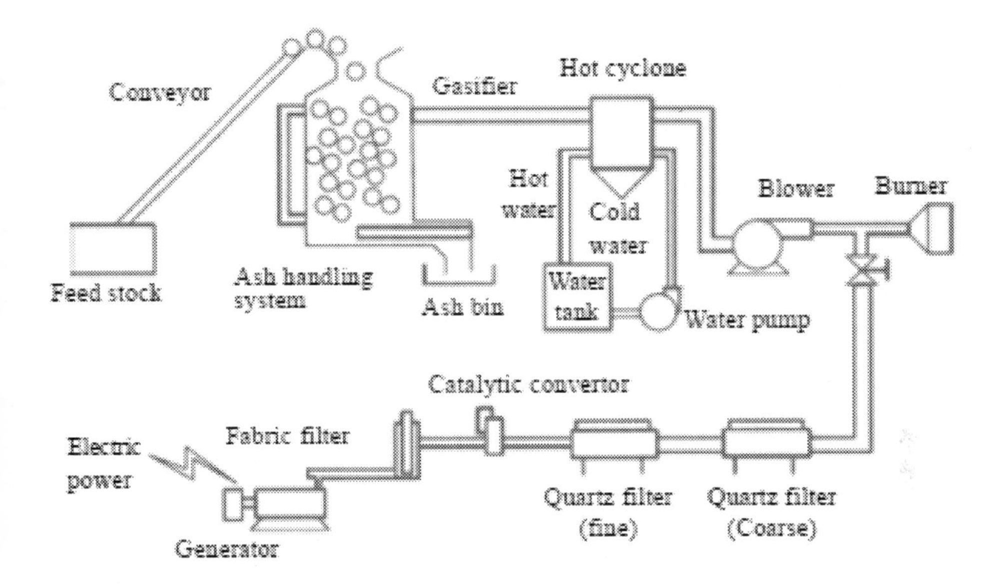

Fig.6.16. Schematic diagram of the gasifier power plant

c. Syngas to methanol and ammonia

In general, the gas produced from biomass must undergo extensive treatment and processing before it is suitable for production of methanol. Tars, acids, sulphur components, and particles have to be removed to obtain a high purity synthesis gas. Fractions of inert gaseous components must be separated and it is often necessary to adjust the ratio of H_2 to CO and CO_2 with catalysts to obtain a gas with the ideal composition for methanol production. The reaction equations are

$$CO + 2H_2 = CH_3OH$$

$$CO_2 + 3H_2 = CH_3OH + H_2O$$

Since methanol production in commercial plants producing synthesis gas from coal have operated for many years this technology is considered to be proven. Several methanol from biomass processes are in the demonstrations phase. If the producer gas is used for ammonia production, a similar extensive downstream processing is necessary. In this case the gas composition must be almost completely shifted from CO to H_2. The reaction equation is :

$$3H_2 + N_2 = 2NH_3$$

Gasification with O_2-enriched air seems appropriate. No plants are planned, probably because of unfavourable economics.

d. Substitute natural gas

A CH_4-rich gas is produced by methanation of the product gas during or after gasification. The reaction equation is:

$$CO + 3H_2 = CH_4 + H_2O$$

Usually catalysis and high pressure are used. If biomass gasification is combined with natural gas reforming, a synthesis gas for methanol production can be produced without the need for a shift reactor.

6.11. Cogeneration

Cogeneration is defined as the sequential production of two different forms of useful energy from a single primary energy source, typically mechanical energy and thermal energy. It mainly consists of the main drive, generator, electrical connections and heat recovery systems.

Mechanical energy can be used to drive an alternator to produce electricity or a rotating device such as a motor, compressor, pump or fan to provide various services. Thermal energy can be used either for direct process applications or for indirect production of steam, hot water, hot air for a dryer or chilled water for process cooling. The efficiency of the cogeneration system is 85% when compared to conventional electricity generation is 58%.

Since the heat and power are simultaneously generated, the $3/4^{th}$ of the primary energy source requirement reduced in the cogeneration process.

6.11.1. Classification of cogeneration systems

Cogeneration systems are usually classified according to the order of energy use and the operating schemes adopted. A cogeneration system can be classified as either a topping or bottoming cycle based on the sequence of energy consumption.

a. Topping cycle

In this cycle, the supplied fuel is used first to produce power and then thermal energy, which is a by-product of the cycle, is used to satisfy process heat or other heat requirements. Cogeneration with a topping cycle is widely used and is the most popular method of cogeneration.

b. Bottoming cycle

In the bottoms cycle, the primary fuel produces high temperature thermal energy and the heat removed from the process is used to generate power through a recovery boiler and turbine generator. Bottom cycle cogeneration is not widely used compared to heating cycle cogeneration.

6.11.2. Types of cogeneration system

1. Steam turbine cogeneration systems
2. Gas turbine cogeneration systems and
3. Reciprocating engine cogeneration systems

6.11.2.1. Steam turbine cogeneration systems

These types of turbine are widely used for the past 100 years. The specific advantage of using steam turbines in comparison with the other prime movers is the option for using a wide variety of conventional as well as alternative fuels such as coal, natural gas, fuel oil and biomass. In the steam turbine, the incoming high pressure steam is expanded to a lower pressure level, converting the thermal energy of high pressure steam to kinetic energy through nozzles and then to mechanical power through rotating blades.

The power generation efficiency of the demand for electricity is greater than one MW up to a few hundreds of MW. Due to the system inertia, their operation is not suitable for sites with intermittent energy demand. Another variant of the cogeneration system with a steam turbine heating cycle is the backpressure extraction turbine, which can be used where the end user needs thermal energy at two different temperature levels. Fully condensing steam turbines are usually installed in places where the heat removed from the process is used to generate power. The two most commonly used types of steam turbines are the back pressure and the extraction condensing turbine.

a. Back pressure turbines

In this type, steam enters the turbine chamber at high pressure and expands to low or medium pressure (Fig. 6.17). The enthalpy difference is used to produce energy/work. Depending on the pressure (or temperature) levels at which the process steam is required, back pressure steam turbines can have different configurations.

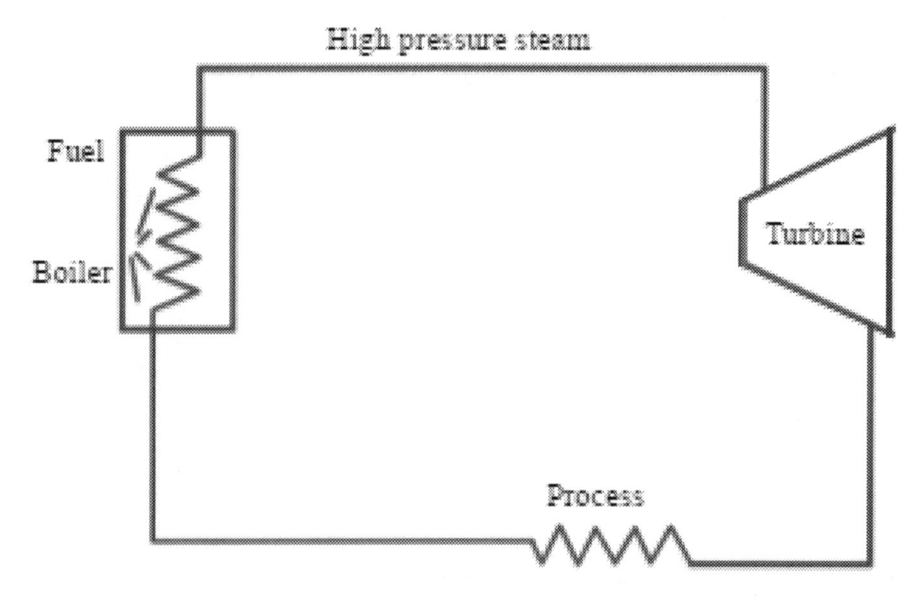

Fig.6.17. Back pressure turbine

b. Extraction condensing turbine

In Extraction cum Condensing steam turbine as shown in Fig.6.18, high pressure steam enters the turbine and exits the turbine chamber gradually. In a two-stage extraction condensing turbine, medium pressure steam and low pressure steam exit to meet the needs of the process. The balance quantity condenses in the surface condenser. In this type, steam entering at high/medium pressure is extracted at medium pressure in the turbine for process use, while the remaining steam continues to expand and condense in the surface condenser and work is done until it reaches condensing pressure (vacuum). The energy difference is used for generating power.

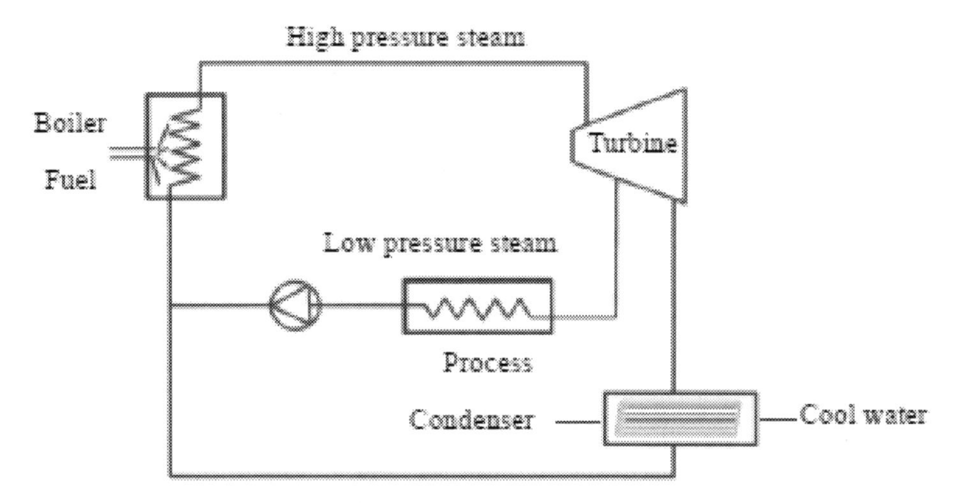

Fig.6.18. Extraction condensing turbine

6.11.2.2. Gas turbine cogeneration systems

Gas turbine cogeneration systems basically work on the principle of braton cycle. Steam generated from gas turbine exhaust passes through a back-pressure or extraction-condensation steam turbine to produce additional power. The waste or extracted steam from the steam turbine provides the necessary thermal energy. It can produce all or a part of the energy requirement of the site, and the energy released at high temperature in the exhaust stack can be recovered for various heating and cooling applications. Though natural gas is most commonly used, other fuels such as light fuel oil or diesel can also be employed. The typical range of gas turbines ranges from a fraction of a MW to around 100 MW. Gas turbine cogeneration has probably experienced the most rapid development in the recent years due to the greater availability of natural gas, rapid progress in the technology, significant reduction in installation costs, and better environmental performance. Furthermore, the gestation period for developing a project is shorter and the equipment can be delivered in a modular manner. Gas turbine has a short start-up time and provides the flexibility of intermittent operation. Although it has a low heat-to-energy efficiency, more heat can be obtained at higher temperatures. If the heat output is lower than what the user requires, supplemental natural gas combustion can be used by mixing additional fuel with the oxygen-rich exhaust gas to increase the heat output more efficiently. It consists of two cycles namely open and closed cycle gas turbines (Fig.6.19a & b).

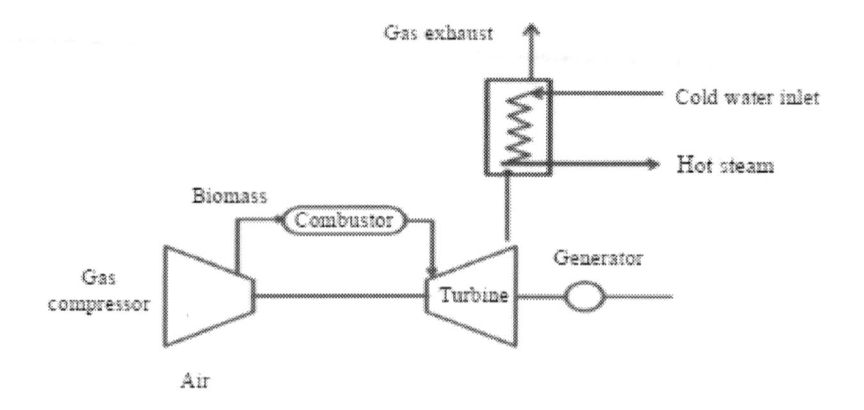

Fig.6.19a. Open cycle gas turbine

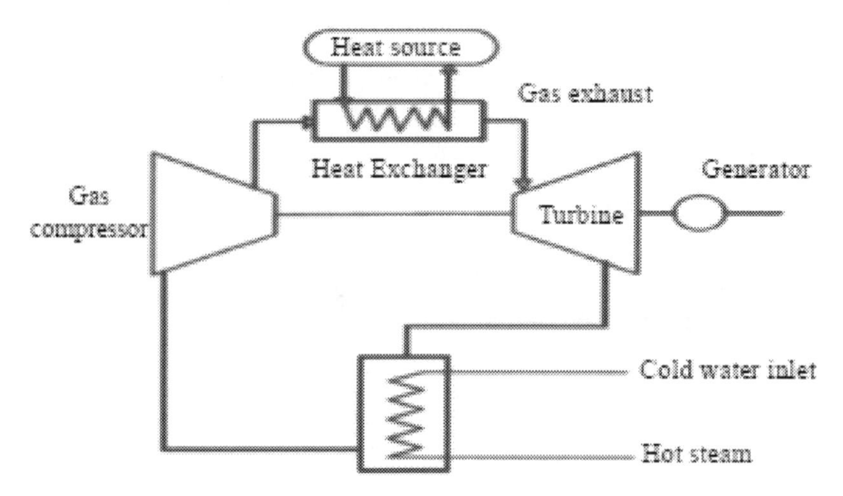

Fig.6.19b. Closed cycle gas turbine

6.11.2.3. Reciprocating engine cogeneration systems

Also known as Internal Combustion (I. C.) engines, these cogeneration systems have high power generation efficiencies in comparison with other prime movers. There are two sources of heat for recovery: exhaust gas at high temperature and engine jacket cooling water system at low temperature (Fig.6.20). As heat recovery can be quite efficient for smaller systems, these systems are more popular for equipment with lower energy consumption, especially those with a greater need for electricity than thermal energy and where the heat quality required is not high, e.g. low pressure steam or hot water.

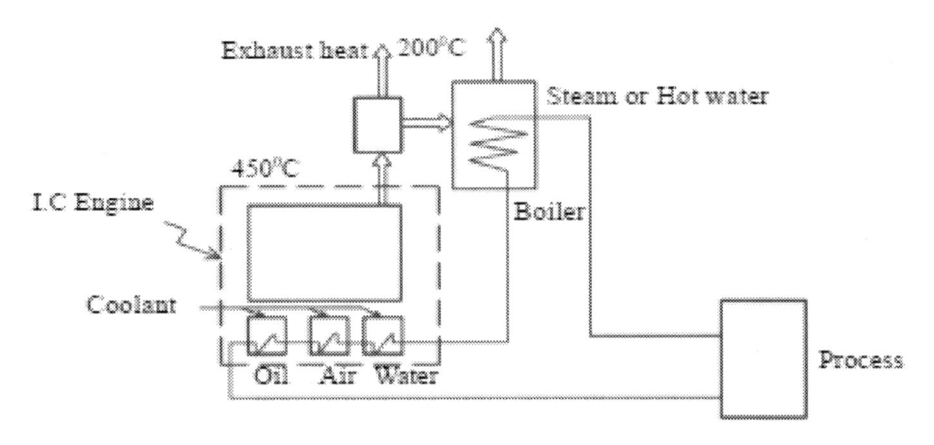

Fig.6.20. Reciprocating gas turbine

6.12. Advantages of Cogeneration

1. Increased efficiency of energy conversion and utilization
2. Lower emissions into the environment, especially CO2, the main greenhouse gas
3. In some cases, biomass fuels and some waste materials such as refinery gases, process or agricultural waste (either anaerobically digested or gasified) are used. These substances, which serve as fuels for cogeneration systems, increase cost efficiency and reduce the need for waste disposal
4. Large cost savings that provide additional competitiveness for industrial and commercial users while offering affordable heat for domestic users
5. The opportunity to move to more decentralized forms of electricity generation, where power plants are designed to meet the needs of local consumers, provide high efficiency, avoid transmission losses and increase flexibility in system use. This will be especially the case when the energy carrier is natural gas
6. An opportunity to increase the variety of generation races and provide competition in generations. Cogeneration represents one of the most important tools for supporting the liberalization of energy markets.

Chapter - 7

Solar Energy

Solar energy is the electromagnetic energy transmitted from the sun (solar radiation). The amount that reaches the earth is equal to one billionth of total solar energy generated, or the equivalent of about 420 trillion kWh. This energy keeps the Earth's temperature above that of a cooler space, causes currents in the atmosphere and ocean, causes the water cycle, and generates photosynthesis in plants. The amount that reaches Earth is equal to one billionth of the total solar energy, equivalent to about 420 trillion kWh. The energy radiated by the sun on a clear sunny day is approximately 1 kW/m2, attempts have been made to harness this energy to raise steam which can be used to drive propulsion machinery to produce electricity. However, due to the large space required, the uncertainty of the availability of power at a constant rate, due to cloud cover, wind fog, etc., the use of this source in generating electricity is limited.

7.1. Characteristics of Solar Radiation

Energy is emitted by the sun as electromagnetic waves, 99% of which have a wavelength between 0.2 and 4.0 micrometers (1 micrometer = 10^{-6} meters). Solar energy reaching the top of the earth's atmosphere consists of about 8% ultraviolet radiation (short wave length, less than 0.39 micrometer), and 46% infrared radiation (long wave length more than 0.78 micrometer).

7.1.1. Solar constant

The sun is a large sphere of very hot gases; the heat being generated by various kinds of fusion reactions. Its diameter is 1.39 x 10^6 km. While that of the earth is 1.27 x 10^4 km. The mean distance between the two is 1.50x10^8 km. Although the sun is large, it subtends an angle of only 32 minutes at the earth's surface. This is because it is also at a very large distance. Thus the beam radiation received from the sun on the earth is almost parallel. The brightness of the sun varies. The rate at which solar energy arrives at the top of the atmosphere is called the solar constant I_{sc}. This is the amount of energy received in unit time on a unit area perpendicular to the sun's direction at the mean distance of the earth from the sun. Because of the sun's distance and activity vary throughout the year; the rate of arrival of solar radiation varies accordingly. It is called solar constant in thus an average from which the actual values vary upto 3% in either direction. This variation is not important however, for Administration's (NASA) standard value for the solar constant, expressed in three common units, is as follows:

1.353 kiowatts per square metre or 1353 watt per square metre (or)

116.5 langleys (calories per sq.cm) per hour (or)

1165 kcal per sq.m per hour (1 langley = 1cal/cm^2 of solar radiation received in 1 day)

7.1.2. Solar radiation at the earth's surface

Solar radiation received at the surface of the earth is entirely different due to the various reasons. The solar radiation that penetrates the earth's atmosphere and aches the surface differs in both amount and character from the radiation at the top of the atmosphere. In the first place, part of the radiation is reflected back into the space, especially by clouds. Furthermore, the radiation entering the atmosphere is partly absorbed by molecules in the air. Oxygen and ozone (O_3), formed from oxygen, absorb nearly all the ultraviolet radiation, and water vapour and carbon dioxide absorb some of the energy in the infrared range. In addition, part of the solar radiation is scattered (i.e., its direction has been changed) by droplets in clouds by atmospheric molecules, and by dust particles

7.1.3. Types of solar radiation

Solar radiation that has not been absorbed or scattered and reaches the ground directly from the sun is called direct ration or Beam radiation. It is the radiation which produces a shadow when interrupted by an opaque object. Diffuse radiation is that solar radiation received from the sun after its direction has been changed by reflection and scattering by the atmosphere. Because of the solar radiation is scatted in all directions in the atmosphere, diffuse radiation comes to the earth from all parts of the sky. The total solar radiation received at any point on the earth's surface is the sum of the direct and diffuse radiation and represented in Fig.7.1. The insolation is defined as the solar radiation energy received on a horizontal surface of unit area (1 m^2) on the ground in unit time (1 day).

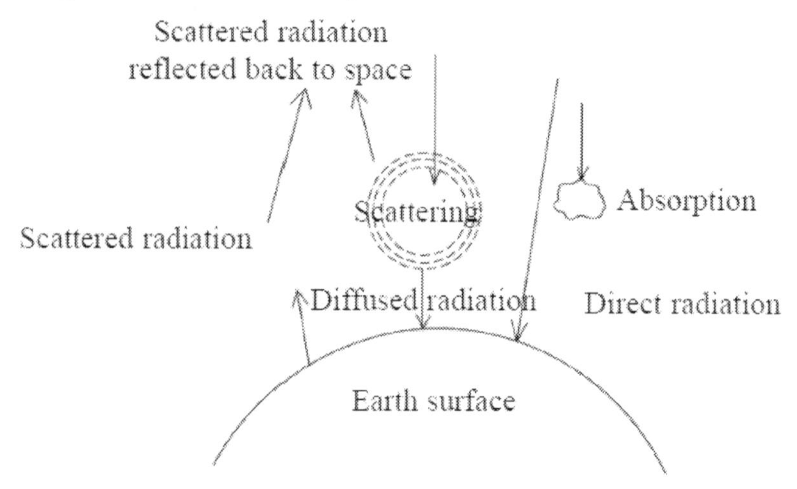

Fig.7.1. Solar radiation

The insolation at a given location on the earth's surface depends, among other factors, on the altitude of the sun in the sky. (The altitude is the angle between the sun's direction and the horizontal). Since the sun's altitude changes with the date and time of the day and with the geographic; attitude at which the observations are made, the rate of arrival of solar radiation on the ground is a variable quantity even in the time. The air mass (m) is the path length of radiation through the atmosphere, considering the vertical path at sea level as unity.

7.1.4. Solar radiation geometry

In solar radiation analysis, the following angles are useful and represented in Fig.7.2.

ϕ_L = latitude of location, δ= declination, ω= hour angle, γ_s = solar Azimuth angle, β = slope, α= altitude angle and θ_z = zenith angle

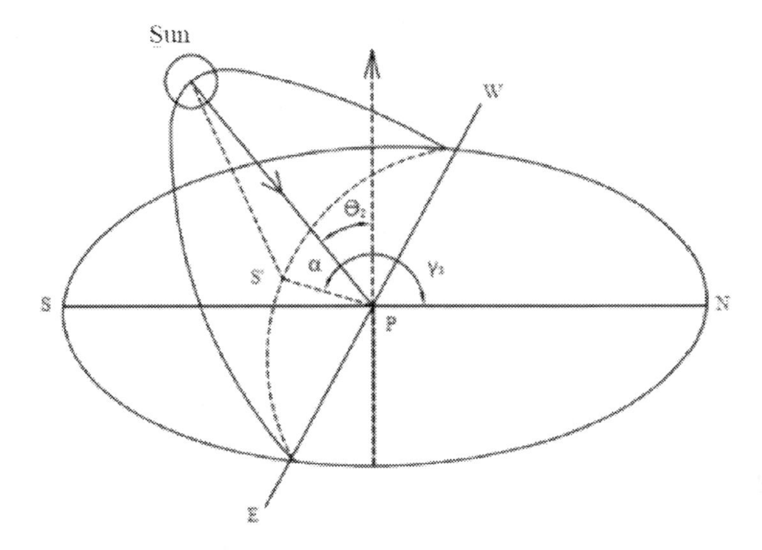

Fig.7.2. Diagram Illustrating the about various angles

a. Latitude of location (φ_L)

Latitude of a location is the angle made by the radial line joining the location to the centre of the earth with the projection of the line on the equatorial plane. The location may be either to the north or south of equator (north is positive, and south is negative) $-90^0 < \varphi_L < 90^0$. It is the angular displacement of a location north or south of equator. The equator is 0°, the North pole is 90° North and the South pole is -90° South.

b. Longitude

The longitude is the angle between prime meridian and the meridian of the place varies from 0 to 180 degree.

c. Solar Declination (δ)

The solar declination is the angle between the equatorial plane and the ecliptic plane. The solar declination angle varies with the season of the year, and ranges between -23.5^o and $+23.5^o$.

d. Zenith Angle (θ_z)

Angle between sun's ray and a line perpendicular to the horizontal plane at P. Zenith is the point in the sky directly overhead a particular location as the Zenith angle θ_z increases, the sun approaches the horizon. It is the angle between the vertical and the line to the sun. Air mass (sometimes called air mass ratio) is equal to the cosine of the zenith angle. The zenith angle is that angle from directly overhead to a line intersecting the sun. The air mass is an indication of the length of the path the solar radiation travels through the atmosphere. An air mass of 1.0 means that the Sun is directly overhead and radiation passes through one atmosphere (thickness).

e. Altitude angle (α)

The altitude angle is the angle between the horizontal and a straight line to the sun. It is the complement of the zenith angle.

f. Azimuth angle (γs)

The Solar Azimuth angle (γ_s) is the angular displacement from south of the projection of the beam radiation on the horizontal surface. Displacements east of south are negative and west of south of positive. The surface azimuth angle (γ) is the deviation of the projection on the horizontal plane of the normal to the surface from the local meridian, with zero due south, east negative, and west positive; $-180° \leq \gamma \leq +180°$.

g. Slope (β)

The slope is the angle between the plane of the surface in question with the horizontal. i.e., $0 \leq \beta \leq 180°$ ($\beta \geq 90°$ means that the surface is downward facing component).

h. Hour angle (ω)

The hour angle is the angular displacement of the sun east or west of the local meridian due to rotation of the earth on its axis at 15° per hour, morning negative and afternoon positive.

i. Angle of incidence

The angle of incidence is the angle between the beam radiation on the surface and the normal to the surface.

j. Tilt angle

The tilt angle is the vertical angle between the horizontal and the array surface.

7.1.5. Pyranometer

The pyranometer is an instrument used for measuring global solar irradiance.

7.1.5. Pyrheliometer

A pyrheliometer is an instrument used to measure direct solar radiation. It uses a 5.7° aperture to transcribe the sun's disk.

7.2. Solar thermal energy

Solar thermal equipment captures and transfers the thermal energy available in sunlight, which can be used to meet heat requirements in different temperature ranges.

Three main temperature ranges are used, they are

- » Low temperature – Hot water - 60°C to 80°C
- » Medium temperature – Drying - 80°C to 140°C
- » High temperature – Cooking & power generation - > 140°C

Solar thermal systems have four main components:

- » Solar collector,
- » Working fluid,
- » Storage tank,
- » Controller.

The heat captured by an active system can be used to heat water or air, or to power a pump. When solar collectors are exposed to sunlight, the working fluid flowing in the passages or tubes in the panels heats up. The fluid is usually water, water and antifreeze or coolant. The transparent cover above the collector slows down the escape of accumulated heat into the outside air. The working fluid is circulated by a pump or fan. The controller turns the system on when the sun's energy is available and off during cloudy period sand at night. The controller can also automatically drain water from the collector to prevent it from freezing and cracking the collector tubes.

7.2.1. Solar energy collectors

A solar collector is a device for collecting solar radiation and to transfer the energy to a fluid passing in contact with it. Utilization of solar energy requires solar collectors.

These are general of two types:

i. Non concentrating or flat plate type solar collector

ii. Concentrating (focusing) type solar collector

A solar energy collector with an associated absorber is an essential part of any system for converting solar energy into a more usable form (eg heat or electricity). In the non-concentrating type, the area of the collector (i.e. the area that captures solar radiation) is the same as the area of the absorber (i.e. the area absorbing the radiation). On the other hand, in the case of concentration collectors, the area capturing solar radiation is larger, sometimes even a hundred times larger, than the area of the absorber. Using concentrating collectors, much higher temperatures can be achieved than with the non-concentrating type. Concentrating collectors can be used to produce medium pressure steam. They use many different arrangements of mirrors and lenses to concentrate the sun's rays on the boiler. This type shows better efficiency than the flat plate type. For best efficiency, collectors should be mounted to face the sun as it moves through the sky.

7.2.1.1. Flat plate type solar collectors

They are particularly suitable where temperatures below approx. 90°C are suitable, such as with spatial and utility collectors for heating utility water, which are of the non-concentrating type. They are produced in rectangular panels with an area of about 1.7 to 2.9 m^2 and their construction and assembly is relatively simple. Flat plates can collect and absorb both direct and diffuse solar radiation; they are consequently partially effective even on cloudy days when there is no direct radiation. Flat solar electrics can be divided into two main classifications according to the type of heat transfer fluid used. Liquid heating collectors are used for heating water and non-freezing aqueous solutions and occasionally for non-aqueous heat transfer fluids. Air or gas heating collectors are used as solar air heaters. The principal difference between the two types is the design of the passage for the heat for the transfer fluid.

The majority of the flat – plate collector has five main components as follows:

1. A transparent cover, which may be one or more panes of glass or plastic sheeting or radiation-transmitting sheeting.

2. The tubes, fins, passage or channels are integral with or connected to the collector absorber plate and convey water, air or other fluid.

3. An absorption plate, usually metal or with a black finish, although a wide variety of other materials can be used with air heaters.

4. Insulation that should be provided at the back and sides to minimize heat loss. Standard insulation materials such as fiberglass or styrofoam are used.

5. A cover or container that surrounds other parts and protects them from the elements.

a). Typical liquid collector

There are many flat plate collector designs, but most are based on the principle shown in Fig.7.3. It is the plate and tube type collector. Basically, it consists of a flat surface with a high absorption of solar radiation, which is called an absorbing surface. The most commonly used materials are usually metal plates, usually of copper, steel or aluminum material with copper tubes in thermal contact with the plates. The absorber plate is usually made of sheet metal with a thickness of 1 to 2 mm, while the tubes, which are also metal, have a diameter of 1 to 1.5 cm. They are brazed, soldered or clamped to the bottom (in some cases the top) of the absorber plate corrugated galvanized sheet is a material widely available worldwide.

Heat is transferred from the absorber plate to the point of use by circulating a fluid (usually water) over the solar heated surface. Thermal insulation with a thickness of 5 to 10 cm is usually placed behind the absorber plate to prevent heat loss from the rear surface. The insulating material is generally mineral wool or glass or glass fibers as mentioned above.

The front covers are usually glass (there may be one or more) which is transparent to incident solar radiation and opaque to infrared back radiation from the absorber. The glass covers act as a conventional shield to reduce losses from the absorber plate below. Glass is generally used for the transparent covers but certain plastic films may be satisfactory. Glass is the most favourable material. Thickness of 3 and 4 mm are commonly used. The usual practice is to have 1 or 2 covers with a specific ranging from 1.5 to 3 cm.

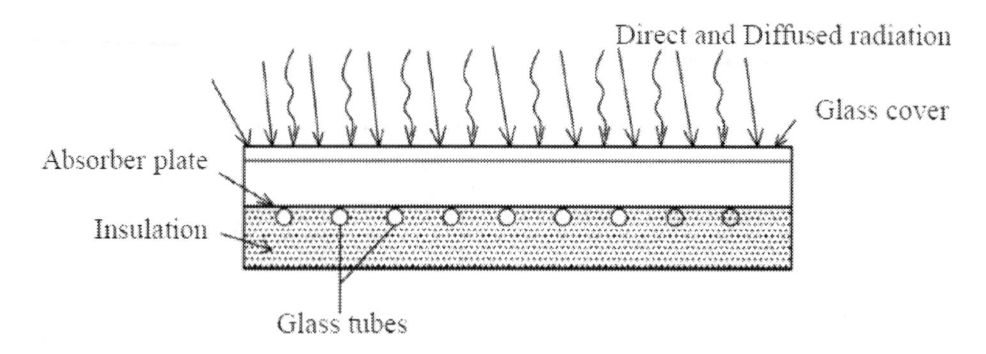

Fig.7.3. Flat plate collector

The advantage of second glass which is added above the first one are: 1. Losses due to air convection are further reduced. This is important in windy areas. 2. Radiation losses in the infra-red spectrum are reduced by a further 25%, because half of the 50% which is emitted outwards from the first glass plate is back radiated. It is not worthwhile to use more than two glass plates. This is due to the fact that each plate reflects about 15% of the incoming sunlight.

b). Heat transport system

The heat generated in the absorber is removed by continuous flow of a heat transport or heat transfer medium, either water or air. Water is a very effective heat transport medium, but it suffers from certain drawbacks, one is the possibility of freezing in the collector tubes in cold climates during cold night. As stated earlier ethylene glycol is added to prevent freezing, but this generally adds to the complexity of the heating system. Furthermore, the antifreeze solution is less effective than water for heat removed from the absorber. In some cases, the water is drained from the collector tubes if freezing is expected, but difficulties have been experienced in refilling all the tubes in the morning. Another problem arises from corrosion of the metal tubes by the water; this aggravated if the water is drained at night thus allowing air to enter. The oxygen in air increases the rate of corrosion of most metals. Corrosion can be minimized by using copper tubing. Aluminium is a less expensive alternative, although periodic chemical treatment of water is desirable.

Finally, leaks in a water (or anti-freeze) circulation system require immediate attention. Fins attached to the plate increase the contact surface. The back side of the collector is heavily insulated with mineral wool or some other material. The most favourable orientation, of a collector, for heating only is facing due south at an inclination angle to the horizontal equal to the latitude plus 150. Air has been used far to a lesser extent as the heat – transport medium in solar collectors, but it may have some advantages over water. To decrease the power required to pump the necessary volume of air through tubes, wider flow channels are used. The solar air heaters, which supply hot air that could be mainly used for heating buildings and green houses, drying agricultural produce and lumber, air conditioning buildings utilizing desiccant beds or an absorption refrigeration process and using air heaters as the heat sources for a heat engine such as a Brayton or Stirling cycle.

Air heaters are basically classified into the following two categories.Basically air heaters are classified in the following two categories.

1. The first type has a non-porous absorber in which the air stream does not flow through the absorber plate. Air may glow above and or behind the absorber plate
2. The second type has a porous absorber that includes slit and expanded metal, transpired honey comb and over lapped glass plate absorber.

c). Advantages of flat plate type solar collectors

1. Both beam and diffuse solar radiation is used
2. Do not require orientation towards the sun.
3. Requires little maintenance.
4. Mechanically simpler than the concentrating reflectors, absorbing surface and orientation devices of focusing collectors.

7.2.1.2. Concentrating (focusing) type solar collector

A focusing collector is a device for collecting solar energy with a high intensity of solar radiation on an energy-absorbing surface. Such collectors generally use an optical system in the form of reflectors or refractors. A focusing collector is a special form of flat plate collector modified by inserting a reflective (or refracting) surface (concentrator) between the solar radiation and the absorber.

These types of collectors can have radiation increases from as low as 1.5–2 to as high as 10,000. As a result of the energy concentration, fluids can be heated to temperatures of 500 °C or more.

a) Line focusing collectors (Trough reflector)

The principle of the parabolic trough collector, which is often used in concentration collectors, is shown by the cross- section in Fig.7.4. solar radiation coming from the particular direction is collected over the area of the reflecting surface and is concentrated at the focus of the parabola, if the reflector is in the form of a through with parabolic cross- section, the solar radiation is focused alone a line. Mostly cylindrical parabolic concentrators are used, in which absorber is placed along focus axis. The collector pipe, preferably with a selective absorber coating, is used as an absorber.

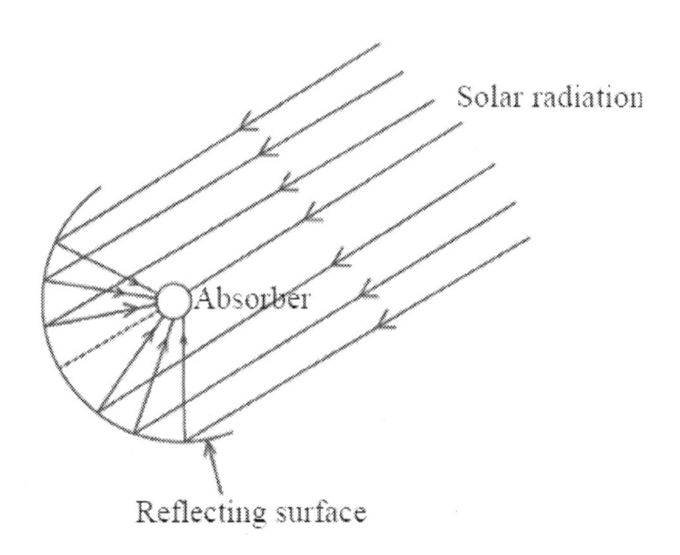

Fig.7.4. Trough type collector

b) Point focusing collector (Paraboloidal type)

A paraboloidal dish collector brings solar radiation to a focus at a point actually a small central volume. A dish 6.6 m in, diameter has been made from about 200 curved mirror segments forming a paraboloidal surface and cross section of a paraboloidal type is shown in Fig.6.5. The absorber, located at the focus, is a cavity made of a zirconium - copper alloy with a black chrome selective coating. The heat transport fluid flows into and out of the absorber cavity through pipes bonded to the interior. The dish can be turned automatically about two axes (up – down and left – right) so that the sun is always kept in a line with the focus and the base (vertex) of the paraboloidal dish. Thus, the sun can be fully tracked at essentially all times.

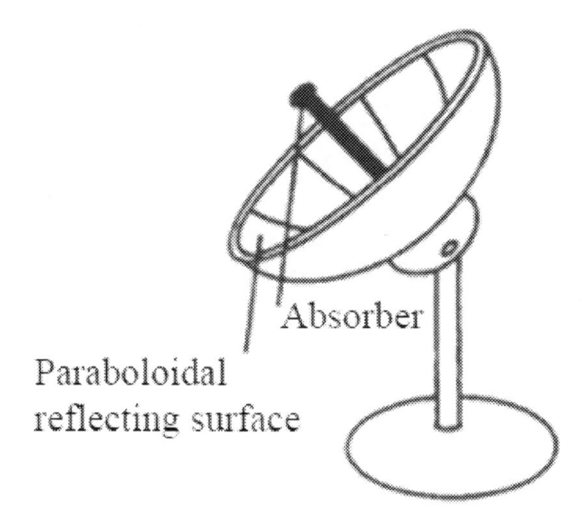

Fig.7.5. Paraboloidal type collector

c). Advantages of concentrating type solar collector

1. Reflecting surfaces required less material and are structurally simpler than flat plate collectors.

2. The absorber area of a concentrator system is smaller than that of flat-plate system for same solar energy collection and therefore the insolation intensity is greater.

3. Because of the area from which heat is lost to the surroundings per unit of the solar energy collecting area is less than that for flat plate collector.

4. Owing to the small area of absorber per unit of solar energy collecting area, selective surface treatment and/or vacuum insulation to reduce heat losses and improve collector efficiency are economically feasible.

5. Focusing or concentrating systems can be used for electric power generation when not used for heating or cooling.

6. Because the temperature attainable with concentrating collector system is higher, the amount of heat which can be stored per unit volume is large and consequently the heat storage costs are less for concentrator systems than for flat plate collectors.

7. Little or no anti-freeze is required to protect the absorber in a concentrator system.

7.2.1.3. Applications of solar thermal systems

a) Solar cookers

Two different types of solar cookers i.e. indirect and direct focusing type have been developed in the country. Such solar cookers are sold on a commercial scale in most states through State Energy Development Corporations or other nodal agencies of the Ministry of Non-Conventional Energy, Government of India.

i) Box type solar cooker

The important components of a solar cooker include an outer box with a heat insulator, an inner cooking box or tray, a double glass lid, a mirror and cooking containers according to Fig. 6.6. The inner cooking box or tray is made from aluminum sheet and coated with black paint to absorb solar radiation and to transfer the heat to the cooking pots. The cooking tray is covered with a double glass lid in which the two glass sheets are spaced at about 20 mm to entrap air which acts a insulator and prevents escape of heat from the inside. The space between the outer box and inner tray including bottom of the tray is packed with insulating material such as glass wool pads to reduce heat losses from the cooker. Cooking containers (with a cover) are generally made of aluminum and painted black on the outer surface to also absorb sunlight directly. The cooking time will depend on the type of food, the time of day and the intensity of sunlight.

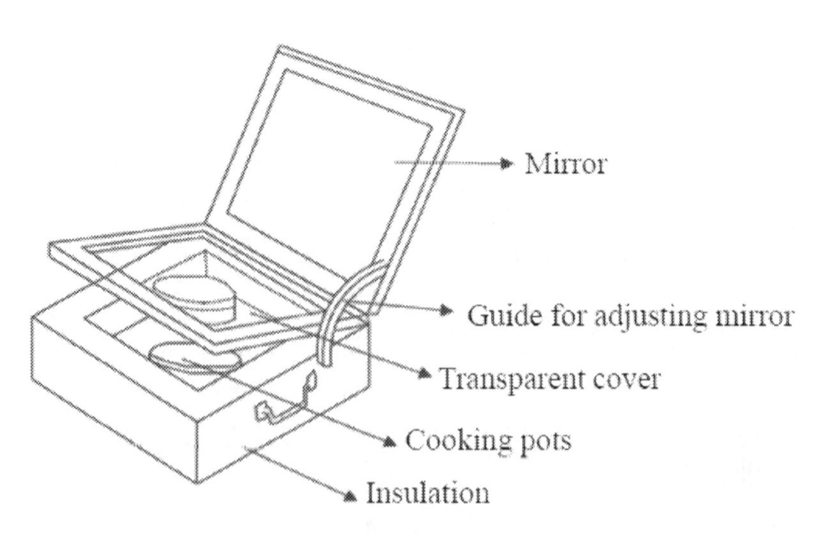

Fig.7.6. Box Type Cooker

b) Solar concentrating cookers

In this type of cookers, the light is concentrated into the cooking vessel through reflectors. The concentration of solar radiation is about 20 to 50 times higher than box type cookers. Therefore, high temperature is obtained for cooking and enables faster cooking.

This can be classified into various types:

a. Dish type cookers
b. Scheffler cookers
c. Parabolic concentrating cookers

i) Dish type cookers

Dish type cooker is a concentrating type parabolic dish cooker with aperture of 1.4 m diameter and focal length 0.28 m. The reflecting material used is anodized aluminium sheet with a reflectivity of over 75%. The parabolic dish is made of reflecting sheets supported on suitable rings for holding them in fixed position. The sheets are joined together in such a way as to form a parabolic shape. The structure and frame of the bowl should be strong so that the reflectors do not get deformed while turning in various directions. The reflector stand is made of mild steel with powder coating. Manual tracking arrangement is provided and adjustments may be made once in 15 to 20 minutes at the time of cooking. The stand is designed in such a way that the reflector could rotate 360° around horizontal axis passing through the focus and centre of gravity. It will also rotate around the vertical axis so as to turn the reflector in the direction of the sun. It is shown in the Fig.7.7.

Fig.7.7. Dish type solar cooker

ii) Scheffler cookers

Scheffler cookers are having concentration ratio, which is more than that of dish type cookers. In this type of cookers heat can be generated at one place and can be transferred to convenient place (like inside a kitchen) as shown in Fig.7.8. In this way, this type of cookers overcomes the inconvenience of cooking under sun. This system can attain the temperature upto 450 °C. An 8 m^2 reflector standing outside the kitchen reflects the sun's rays into the kitchen through an opening in the north wall, while a secondary reflector further focuses the rays onto the bottom of the pot/ pan, which is painted black to absorb the heat. faster and better.

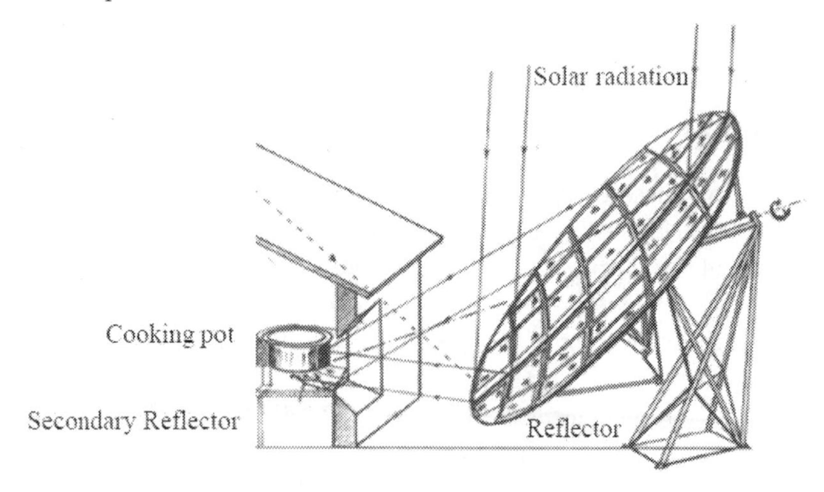

Fig.7.8. Scheffler type cooker

iii) Parabolic concentrating cookers

The solar parabolic cooker works on the principle of concentrating solar energy using a reflective parabolic solar concentrator. A high temperature is achieved because the sun's rays are concentrated on the vessel. General handling is easy and simple. The system uses automatic tracking solar concentrators that convert water into high-pressure steam and consists of a shell receiver made of 35 cm diameter mild steel, connected to two concentrators on either side that focus sunlight on either side of the receiver. The system is shown in Fig. 7.9.

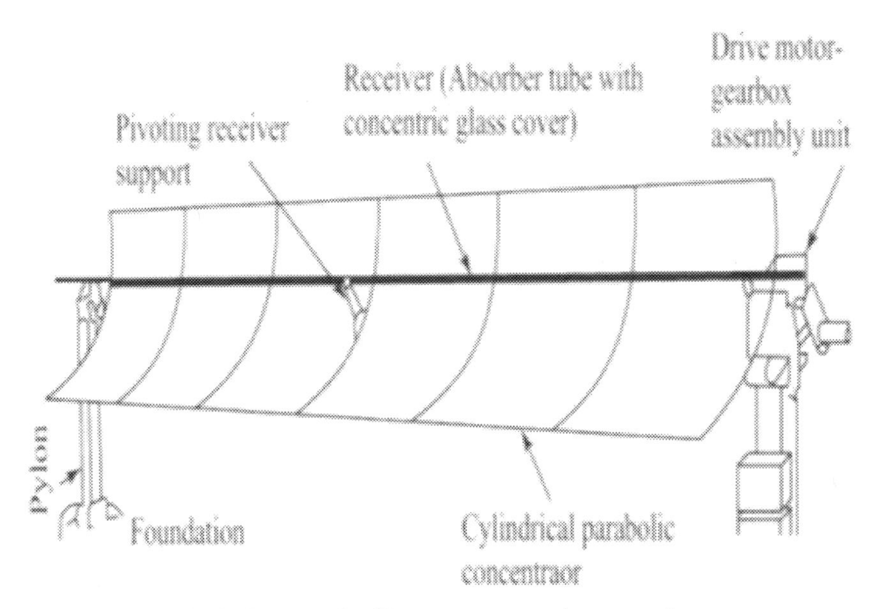

Fig.7.9. Parabolic concentrating cooker

c) Solar water heaters

A solar water heating system (SWHS) is a device that supplies hot water at a temperature of 60°C to 80°C using only solar thermal energy without additional fuel. It has three main components, a solar collector, an insulated hot water tank and a cold water tank with the required insulated hot water piping and fittings.

In smaller systems (100-2000 liters per day), hot water reaches the user through natural (thermo-siphon) circulation, for which the storage tank is placed above the collectors. For systems with a higher capacity, a pump can be used for forced water circulation. The two types water heating system are flat plate solar water heaters and evacuated tube solar water heaters

i) Flat plate solar water heaters

A flat plate solar water heating system is used for domestic and industrial applications for heating water upto 60 to 70 °C. A flat plate collector is simple in design and consists of flat collector or absorber, glass cover, insulation, pipes and storage tanks as shown in Fig.7.10.

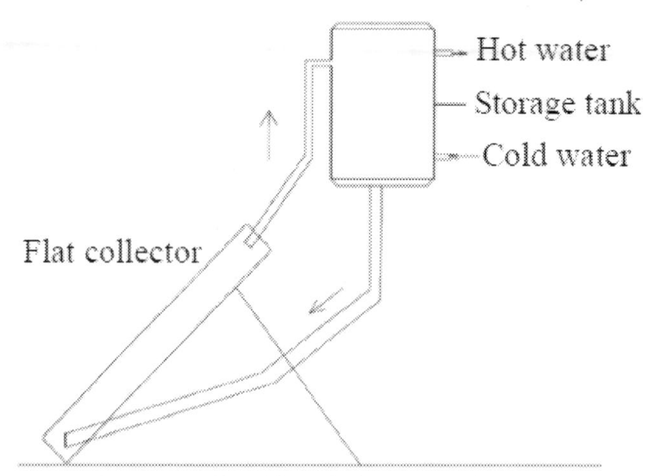

Fig.7.10. Flat Plate Solar Water Heater

ii) Evacuated tube solar water heaters

Evacuated Tube Collector is made of double layer borosilicate glass tubes evacuated for providing insulation as shown in Fig.7.11. The outer wall of the inner tube is coated with a selective absorbent material. This aids in the absorption of sunlight and transfers heat to the water that flows through the inner tube. Due to vacuum in the evacuated tube collector, there is no medium which take the heat away from the absorber plate. Therefore, there are less heat losses. Thus, in evacuated tube collectors, the water temperature close to 100°C can be obtained.

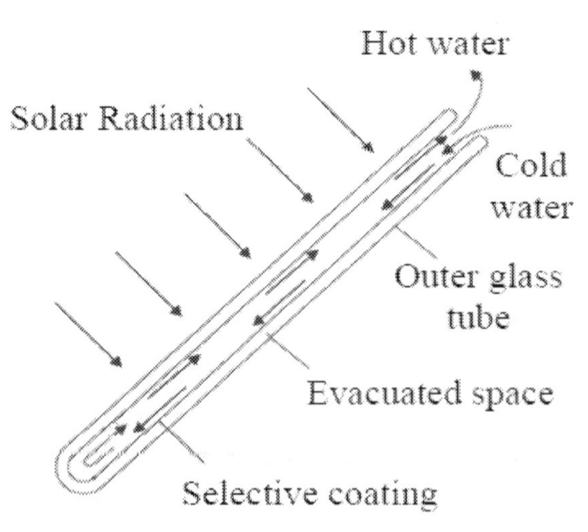

Fig.7.11. Evacuated tube solar water heater

7.2.2. Solar dryer

A solar dryer is an enclosed unit, that protects food from damage, birds, insects and unexpected rain. The rationale behind solar dryers is that they can be more efficient than sun drying, but have lower operating costs than mechanical dryers. Solar dryers utilize the available amount of solar to cover the heat requirements of small dryers during daytime (sunshine hours) operation. The solar dryers are generally classified into two basic categories as natural and forced convection dryers.

a. Natural convection dryers

The natural convection dryers are based on the principle of temperature difference and consequently the difference in the density of the air inside and outside the drying chamber as shown in Fig.7.12. This difference provides the necessary driving force for the air to flow through the drying chamber. Natural-convection dryers do not require any mechanical or electrical power to run a fan. In general, the construction is simple, easy to maintain and inexpensive. On the other hand, the working mechanism is strongly dependent on the temperature difference and the pressure drop that occurs when the air is forced through the product. Therefore, the successful use of natural convection dryers is limited to drying small batches in areas with high solar radiation.

b. Forced convection dryers

To ensure continuous ventilation, forced-convection solar dryers have been introduced. Forced-convection solar dryers generally show several advantages such as high reliability. They require a blower to force air through or over the product and schematic representation of forced convection dryer is given in Fig.7.13. In contrast to natural convection dryers, these dryers are more expensive, complex in construction and are dependent on electrical power. Hence the electricity costs have to be compensated with reduction in drying time, higher drying capacity, reduction of mass losses and better quality of the product.

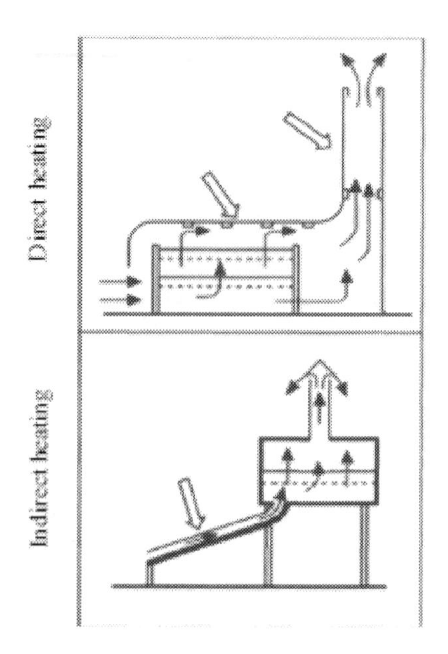

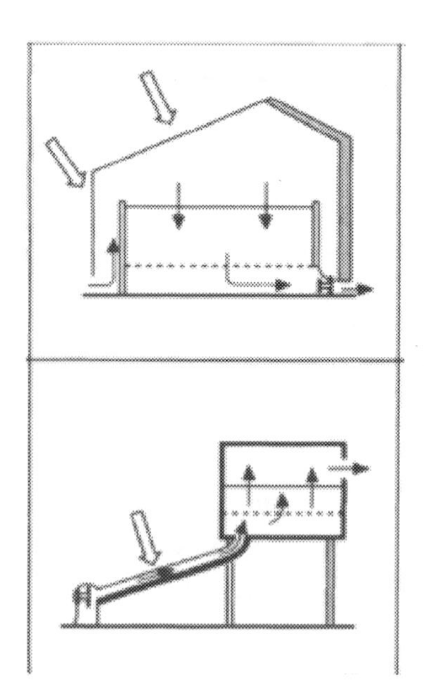

Fig.7.12. Natural convection dryer **Fig.7.13. Forced convection dryer**

c. Solar tunnel dryer

The solar tunnel dryer is commercially used dryer for drying agro food products. The solar tunnel dryer mainly consists of a semi cylindrical pipe structure covered with U.V. stabilized Polyethylene sheet as shown in Fig.7.14. The solar radiation is transmitted through the UV sheet, which has the property of transferring short-wave radiations and opaque for long-wave radiations. When solar radiation in the form of short-wave radiation transmitted through the UV sheet, it gets converted to long-wave radiation inside the solar tunnel dryer. Therefore, the radiation is trapped inside the dryer and increases the temperature of the dryer. For small scale applications natural convection method is preferred and for large scale application the solar tunnel dryer can be combined with blowers to supply hot air in the night time as forced convection dryer

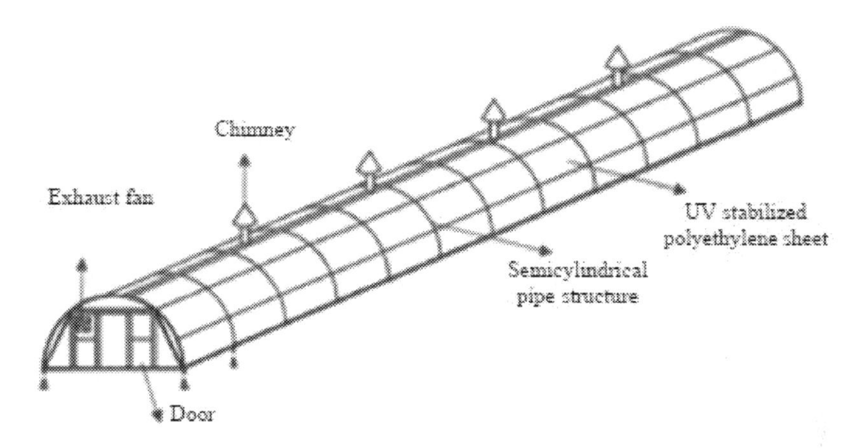

Fig.7.14. Solar tunnel dryer

d. Advantages of solar dryers

1. Solar dryers are more economical compared to dryers that run on conventional fuel/electricity.
2. The drying process is completed in the most hygienic and eco-friendly way.
3. Solar drying systems have low operating and maintenance costs.
4. Solar dryers last longer. A typical dryer will last 15-20 years with minimal maintenance.

e. Disadvantages of solar dryers

1. Drying can only be done during sunny days unless the system is integrated with a conventional energy-based system.
2. Due to the limitations of solar energy collection, the solar drying process is slow compared to dryers that use conventional fuels.
3. Normally, solar dryers can only be used for drying at 40-50°C.

7.2.3. Solar photovoltaic (PV) systems

Photovoltaics (PV) are arrays of cells containing solar photovoltaic material (semiconductor devices) that convert sunlight directly into electricity. A material or device that is capable of converting the energy contained in photons of light into an electrical voltage and current is said to be photovoltaic. A solar cell made of silicon materials, when exposed to sunlight generates voltage and current at its output terminal. Solar cells are connected in series and parallel combinations to form modules that provide required power. The amount of electricity generated depends on the amount of sunlight available to it. The higher the intensity of sunlight, more the electricity generated from it. Photovoltaic systems sometimes have two additional components to complement the solar modules: an inverter and a storage device. Since solar cells produce direct current (DC) and most conventional equipment operates on alternating current (AC), an inverter is used to change the DC current to AC current. The schematic representation of solar PV system with components is given in Fig.7.15.

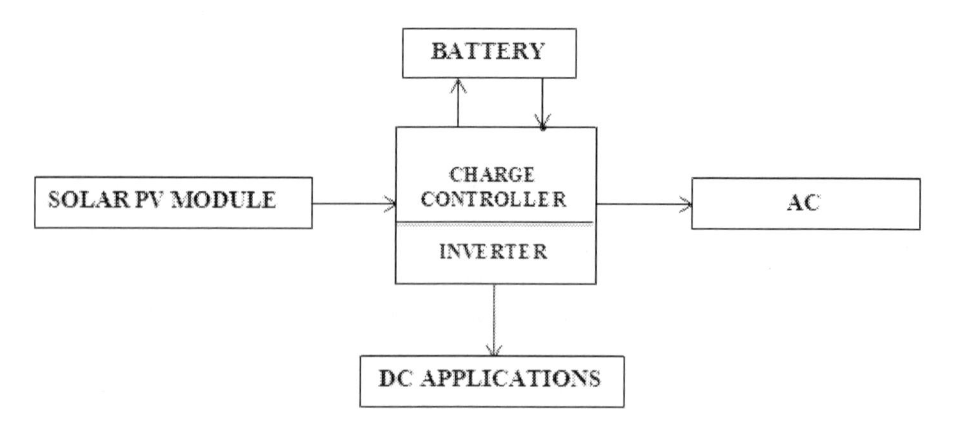

Fig.7.15. Components of Solar PV system

Solar PV module produce power output proportional to their size. Normally solar PV modules are available in the sizes of 5 to 300 W. Since typical silicon solar cells produce only about 0.5 V, many cells are connected together to give higher and useful voltages. Usually 30–36 cells are connected together in a circuit to give a final voltage of about 15 to 17 V, which is sufficient to charge a 12 V battery. If the voltage or current from one module is not enough to power the load, then modules can also be connected together, just as the cells. These modules can be combined together in parallel or series to get required output. Typical solar cells are 3-6 inches in diameter and are available in various shapes such as circular, square etc.

7.2.3.1. Working principle of solar cell

Sunlight can be converted to electricity due to the photovoltaic effect discovered in 1839 by Edmund Becquerel, a French scientist. Sunlight is composed of photons, or packets of energy. When photons strike a solar cell (technically a semiconductor p-n junction device), they get absorbed in solar cell resulting in generation of voltage across the solar cell which can deliver current and thus power to run an electric load. (i.e. the generation of an emf as a result of the absorption of ionizing radiation). The working principle of a solar cell is given in Fig.7.16. In a crystal, the bonds [between silicon atoms] are formed by electrons that are shared by all the atoms in the crystal. The light is absorbed and one of the electrons that is in one of the bonds is excited to a higher energy level and can move more freely than when it was bound. That electron can then move around the crystal freely, and we can get a current.

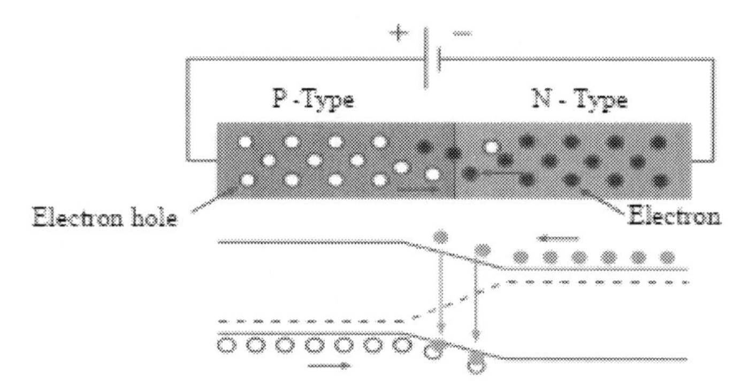

Fig.7.16. Working principle of solar cell

The Table 7.1. summarizes the various types of solar cells available and Fig.7.17. represents the types of solar PV cell

Table 7.1 Types of solar cell and its efficiency

Cell type	Efficiency of cell	Land per MW
Mono crystalline silicon	Around 18 – 24%	3 – 4 acres
Poly/Multi crystalline silicon	Around 14 – 18%	4 – 5 acres
Thin film	Amorphous silicon 6 – 10% Cadmium Telluride 10 – 11% Copper Indium Gallium Diselenide 12 – 14%	7.5 – 9 acres

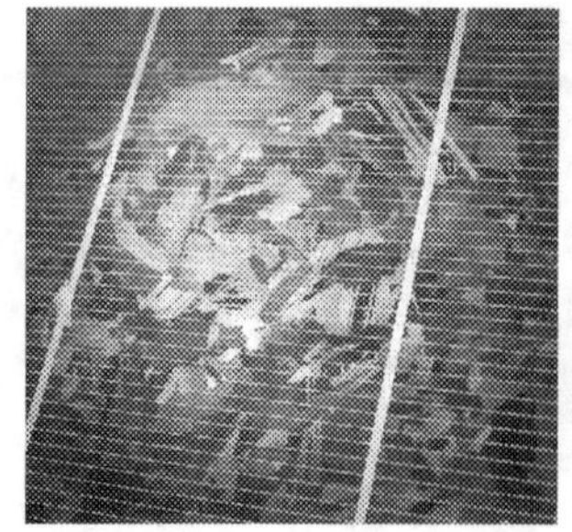
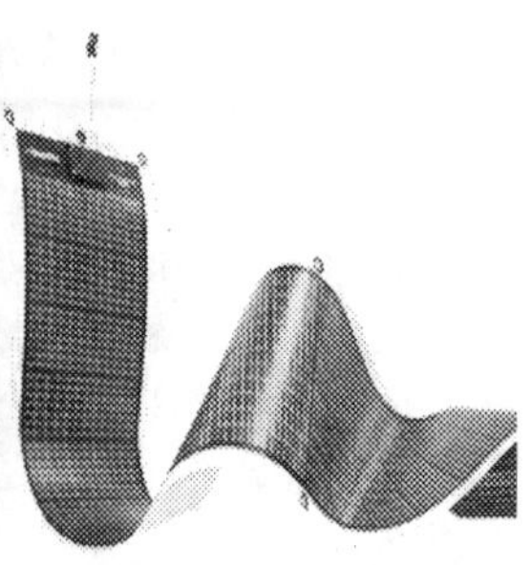

a. Mono crystalline silicon **b. Multi crystalline silicon** **c. Thin film**

Fig.7.17. Types of solar PV cell

7.2.3.2. Applications of solar PV system

a) Solar PV home lighting system

Stand-alone system itself supplies all the energy requirement of the building (for which the system is designed) and does not interact with grid. It is also possible to design a solar PV system that fulfills a part of total energy requirement using solar PV modules and remaining part from the grid. The typical PV system consists of a solar PV module, battery, inverter, charge controller and DC & AC load.

This is a fixed indoor lighting system and consists of solar PV, module, battery and balance of systems.

b) Solar pump

A solar powered pump is a normal pump with an electric motor. Electricity for the motor is generated on site through a solar panel that converts the sun's energy into DC electricity. Because the nature of the electrical output from a solar panel is DC, a solar pump requires a DC motor if it is to operate without additional electrical components. If a pump has an AC motor, an inverter would be required to convert the DC electricity produced by the solar panels to AC electricity. Due to the increased complexity and cost, and the reduced efficiency of an AC system, most solar-powered pumps have DC motors.

Fig. 7.18 shows the components of a solar pump system. The components that make up a solar powered pump depend on whether the pumping system is a direct drive system or a battery powered system. Solar pumping systems can be configured to meet a wide range of requirements. The amount of water a solar powered pump can deliver depends on how far the water needs to be lifted, the distance it has to travel through the discharge pipe (and the size of the pipe), the efficiency of the pump used, and how much power is available to the system. The output can be increased by adding additional solar panels. For example, h.p. a pump powered by three 50-watt panels can deliver water at about 1½ gpm to a location 100 feet above the source. In comparison, the 3 h.p. a low-lift irrigation pump powered by an array of 60 panels could deliver water at about 1,000 gpm

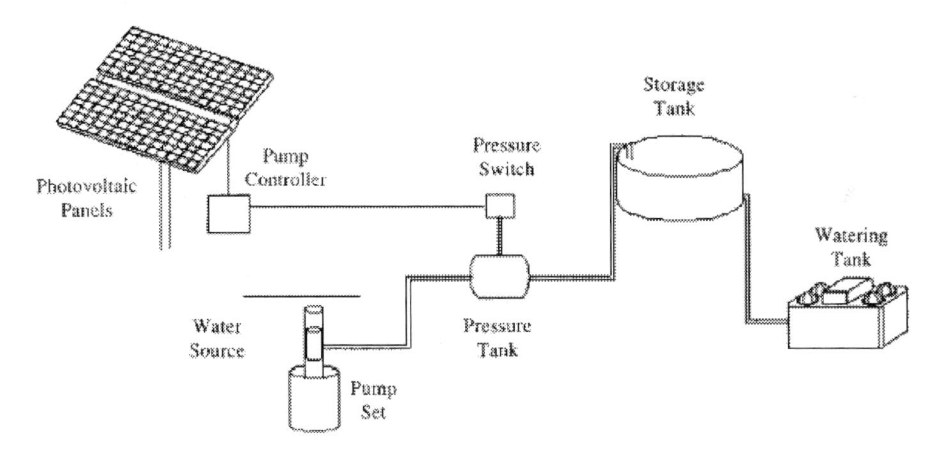

Fig.7.18. Components of solar powered pumping system

i. Direct drive system

A direct drive system where solar panels collect the sun's energy and provide power directly to pumps to pump water.

ii. Battery drive system

A battery operated system where the solar panels collect the suns energy and provides the power to the batteries. The batteries then store the power for use by the pumps at a later date. For direct drive systems, it is important to match the output of the solar array with the pump output requirements to maximize efficiency.

iii. Advantages

» SPV water pumping systems can be installed on site completely eliminating energy losses during transmission.

» Can be installed to the required load of pumps up to 3000W

» SPV modules need only minimum maintenance and no battery is required

» Panels can be cleaned with plain water and a soft cloth. The frequency of inspection should match the amount of storage available

c) Solar street light

The solar streetlight consists of two photovoltaic modules of 36 Watts each mounted on a 6 metre lamppost for charging. At the base of the pole a box is provided which houses the charging system, a storage battery and inverter unit. The unit is also provided with a light sensitive switch so that the street lamp gets lighted after sunset. The panel captures sunlight during daytime and sores in the battery by using a charge controller. The Fig.7.19 represents the components of solar street light system

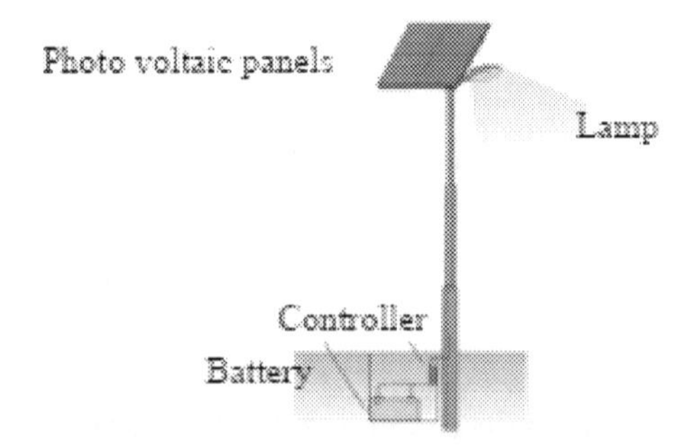

Fig.7.19. Solar Street light

d) Solar lantern

SPV lantern is a portable lamp as shown in Fig.7.20. It consists of SPV module of 10 Wp capacity, rechargeable battery, Compact Fluorescent lamp (CFL) of 5 / 7 W and electronics (i.e. inverter and charge controller). Whenever light is required the DC is inverted and fed to a compact florescent lamp.

Fig. 7.20. Solar Lantern

e) Solar PV power plant

Grid-interactive solar power plants can also be installed to support voltage in rural areas of the grid, reduce peak times in urban centers and save diesel in islands or remote locations. The SPV power plant can also be connected to the grid and the generated electricity could be exported to the grid. PCU senses the grid voltage frequency and modifies the PV array output voltage and frequency so as to synchronize with the grid and feeds the AC power generated from PV array to the grid.

f) Solar PV fencing

Solar PV fencing is used as a fence to avoid animals entering into the farming field and houses in rural areas. The electricity produced from PV cell is stored in batteries and through inverters electricity is supplied to the fences. Once the animal or human being touches the fence an alarm will ring and lose the memory for fraction of a minute and so tress passing is avoided to safe guard the farm field from animals. It costs about Rs.1 lakh per km or Rs.20,000 per acre. The Fig.7.21 represents the working of Solar PV fencing.

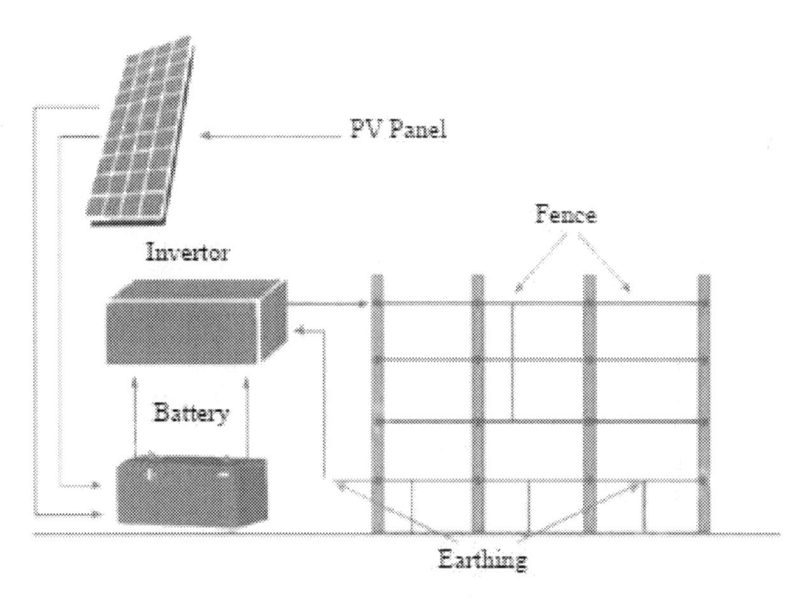

Fig.7.21. Solar Fencing

g) Other advanced applications:

» Solar traps/pest catchers - It costs about Rs. 3,500 / each.

» Solar traffic signals

» Police wireless system in remote areas

» Telecommunication network

» Solar TV and other devices

7.2.3.3. Advantages of photovoltaic solar energy conversion

1. Direct room temperature conversion of light to electricity through a simple solid state device.
2. Absence of moving parts.
3. Ability to function unattended for long periods as evidence in space programme.
4. Modular nature in which desired current, voltages and power levels can be achieved by mere integration.
5. Maintenance cost is low as they are easy to operate.
6. Do not create pollution.
7. Long effective life and highly reliable.
8. They consume no fuel to operate as the sun's energy is free.
9. Rapid response in output to input radiation Changes; no long – time constant is involved, as on thermal systems, before steady state is reached.
10. Wide power handling capabilities from microwatts to kilowatts or even megawatts when modules are combination with power conditioning circuitry to feed power into utility grid.
11. Easy to fabricate, being one of the simplest of semi-conductor devices.
12. Used with or without sun tracking, making possible a wide range of application possibilities.

7.2.4. Solar refrigeration system

Refrigeration is a process of pumping of heat from a high temperature source to a lower temperature sink. From a sustainability perspective, directly using solar as a primary energy source is attractive because of its universal availability, low environmental impact, and low or no ongoing fuel cost. Research has demonstrated that solar energy is an ideal source for low temperature heating applications such as space and domestic hot water heating. Solar heating applications are intuitive since, when solar energy is absorbed on a surface, the surface temperature rises, providing a heating potential. Solar absorption cooling is one of the investigated applications due to the advantages of its use in sunny and warm areas where solar energy can be used as the main source for its operation. Unlike mechanical vapor compression refrigerators, these systems do not deplete the ozone layer and reduce electricity supply requirements. In addition, heat-powered systems could be better than electricity-powered systems because they use cheap waste heat, solar, biomass, or geothermal energy sources, which in many cases have negligible supply costs.

7.2.4.1. Photovoltaic operated refrigeration cycle

Photovoltaics involve the direct conversion of solar radiation to direct current electricity using semiconducting materials. In concept, the operation of a PV-powered solar refrigeration cycle is simple. Solar photovoltaic panels produce dc electrical power that can be used to operate a DC motor, which is coupled to the compressor of a vapor compression refrigeration system. The main considerations in designing a PV refrigeration cycle include appropriately matching the electrical characteristics of the motor driving the compressor with the available current and voltage produced by the PV array. The rate of electricity that a PV system can generate is usually provided by PV module manufacturers for standard rated conditions, i.e. incident solar radiation of 1000 W/m2 (10800 W/ft2) and module temperature of 25°. C (77°F).

Unfortunately, PV modules will operate over a wide range of conditions that are rarely as favorable as the rating condition. Furthermore, the power produced by a PV array is as variable as the solar source from which it is derived. The performance of the PV module, expressed by its current voltage and the power-voltage characteristic, depends mainly on the solar radiation and the temperature of the module. The PV panels can then be selected so that their maximum output voltage is close to the voltage for the battery system. The battery also provides electrical storage so the system can operate when sunlight is not available.

However, the addition of a battery increases the weight of the system and reduces its steady-state efficiency. Electrical storage may not be needed in a solar refrigeration system as thermal storage, e.g., ice or other low temperature phase storage medium, may be more efficient and less expensive.

7.2.4.2. Solar mechanical refrigeration

Solar mechanical refrigeration uses a conventional vapor compression system driven by mechanical power that is produced with a solar-driven heat power cycle. The heat power cycle usually considered for this application is a Rankine cycle in which a fluid is vaporized at an elevated pressure by heat exchange with a fluid heated by solar collectors. A storage tank can be included to provide some high temperature thermal storage. The vapor flows through a turbine or piston expander to produce mechanical power, as shown in Fig.7.22. The fluid exiting the expander is condensed and pumped back to the boiler pressure where it is again vaporized. The efficiency of a solar collector, however, decreases with increasing temperature of the delivered energy. High temperatures can be obtained from concentrating solar collectors that track the sun's position in one or two dimensions. Tracking systems add cost, weight and complexity to the system. If tracking is to be avoided, evacuated tubular, compound parabolic or advanced multi-cover flat plate collectors can be used to produce fluid temperatures ranging between 100°C – 200°C (212°F – 392°F).

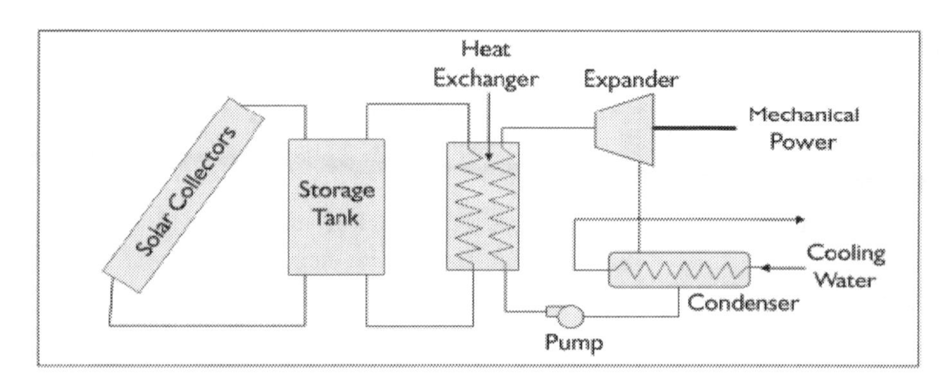

Fig.7.22. Solar driven mechanical power cycle

The efficiency of solar collectors depends on both solar radiation and the difference in temperature between the entering fluid and ambient. The overall efficiency of solar mechanical refrigeration, defined as the ratio of mechanical energy produced to the incident solar radiation, is the product of the efficiencies of the solar collector and the power cycle. Because of the competing effects with temperature, there is an optimum efficiency at any solar radiation. This efficiency is significantly lower than that which can be achieved with non-concentrating PV modules. Solar mechanical systems are only competitive at higher temperatures for which tracking solar collectors are required. Due to economies of scale, this option would only be applicable for large cooling systems (eg 1,000 tons or 3,517 kWT).

7.2.4.3. Absorption refrigeration

Absorption refrigeration is the least intuitive of all solar cooling alternatives. Unlike PV and solar mechanical cooling options, an absorption refrigeration system is considered a thermally driven system that requires minimal mechanical energy for the compression process. It replaces the energy-intensive compression in a vapor compression system with a heat activated "thermal compression system." A schematic of a single-stage absorption system using ammonia as the refrigerant and ammonia-water as the absorbent is shown in Fig.7.23. Absorption cooling systems that use lithium bromide-water absorption- refrigerant working fluids cannot be used at temperatures below 0°C (32°F). The condenser, throttle and evaporator operate in the exactly the same manner as for the vapor compression system.

In place of the compressor, however, the absorption system uses a series of three heat exchangers (absorber, regenerating intermediate heat exchanger and a generator) and a small solution pump. Ammonia vapor exiting the evaporator is absorbed in a liquid solution of water-ammonia in the absorber. The absorption of ammonia vapor into the water-ammonia solution is analogous to a condensation process.

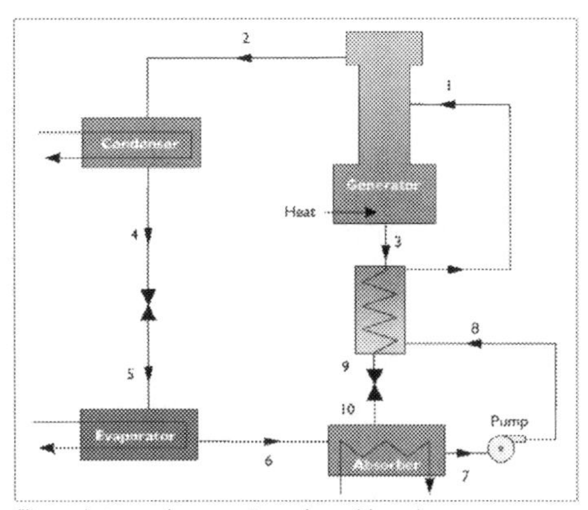

Fig.7.23. Ammonia-water absorption refrigeration system

The process is exothermic and so cooling water is required to carry away the heat of absorption. The principle governing this phase of the operation is that a vapor is more readily absorbed into a liquid solution as the temperature of the liquid solution is reduced. The ammonia-rich liquid solution leaving the absorber is pumped to a higher pressure, passed through a heat exchanger and delivered to the generator. In the generator, the liquid solution is heated, which promotes desorption of the refrigerant (ammonia) from the solution. Unfortunately, some water also is desorbed with the ammonia, and it must be separated from the ammonia using the rectifier. Without the use of a rectifier, water exits with the ammonia and travels to the evaporator, where it increases the temperature at which refrigeration can be provided. This solution temperature required to control the ammonia-water desorption process is in the range between 120°C to 130°C (248°F to 266°F). Temperatures in this range can be obtained using inexpensive non-tracking solar collectors. At these temperatures, evacuated tubular collectors may be more suitable than flat-plate collectors as their efficiency is less sensitive to operating temperature. The overall efficiency of a solar refrigeration system is the product of the solar collection efficiency and the coefficient of performance of the absorption system.

The COP for a single-stage ammonia-water system depends on the evaporator and condenser temperatures. The COP for providing refrigeration at –10°C (14°F) with a 35°C (95°F) condensing temperature is approximately 0.50. The absorption cycle will operate with lower temperatures of thermal energy supplied from the solar collectors with little penalty to the COP, although the capacity will be significantly reduced. The total system coefficient of performance (COP) can be defined as the ratio of the cooling output to the input solar energy.

7.2.4.4. Disadvantages of solar refrigeration

i. Solar refrigeration systems are necessarily more complicated, costly and bulky than conventional vapor compression systems due to the need to locally generate the energy required to operate the refrigeration cycle.

ii. The ability of a solar refrigeration system to function is driven by the availability of solar radiation. Because this energy resource is variable, some form of redundancy or energy storage (electrical or thermal) is required for most applications, which further adds to the system size and cost.

7.2.4.5. Advantages of solar refrigeration

i. The advantage of solar refrigeration systems is that they displace some or all of the conventional fuel use.

ii. The main advantage of solar cooling is that it can be designed to operate independently of the grid.

iii. iii. A photovoltaic system is best suited for small-capacity portable systems located in areas outside of conventional energy sources (electricity or gas).

7.2.4.6. Applications of refrigeration

i) Industrial applications:

a. Processing of food products, farm crops, textiles, photographic materials, tobacco, petroleum and other chemical products.

b. Cooling of concrete for dams.

c. Treatment of air for blast furnace.

ii) Preservation of perishable goods

a. Manufacturing of ice.

b. Freezing or chilling, storage and transportation of food stuffs including beverages, meat, poultry products, dairy products, fish, fruits, vegetables, fruit juices, etc.

c. Preservation of photographic films, archeological documents etc.

iii) Providing comfortable environment

a. Industrial air-conditioning.

b. Comfort air conditioning for hospitals, residences, hotels, restaurants, theaters, offices, etc.

7.2.5. Solar pond

In order to reduce the cost of large solar thermal installations, it is necessary to devise more economical methods of collecting and storing solar energy. In this context, attention has been focused on the possibility of using large expanses of water of small depth for absorbing and storing solar radiation instead of using flat plate collectors and hot water storage tanks. One would obtain a significant rise in the water temperature only if the convection could be prevented. A man-made pond in which there is a significant increase in temperature in the lower regions by preventing convection is called a solar pond. A common method used to prevent convection is to dissolve salt in water and maintain a concentration gradient. For such ponds, the more specific term salt gradient solar pond is used. The working principle of solar pond is illustrated in Fig.7.24.

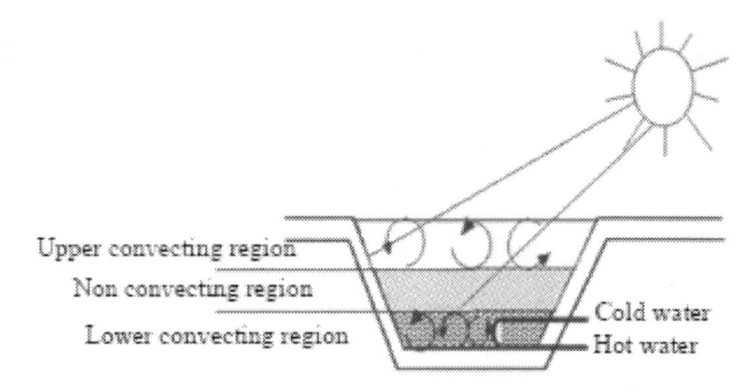

Fig.7.24. Solar pond

It appears that the least expensive type of solar collector is a large solar pond. However, they were mainly considered for large industrial applications, as their price decreases considerably with increasing size. There are two basic types of solar ponds: a shallow pond and a salt gradient pond. Both are equivalent to horizontal flat plate collectors because they are non-concentrating, receive both direct and diffuse solar energy, and are limited to low-temperature applications. One square meter of pond surface can supply 0.5 to 2 GJ of thermal energy per year at a temperature of 40 to 80 °C

7.2.5.1. Salient features of a solar pond

» A body of water that collects and stores solar energy.

» Heated water – rises as it becomes less dense.

» Water loses its heat to the air by convection or evaporates, taking the heat with it.

» Cold water moves down to a natural convective circulation that mixes the water and dissipates the heat.

» Reduces either convection or evaporation in order to store the heat collected by the pond. They can operate in almost any climate.

7.2.5.1. Classification of solar ponds

The solar ponds are mainly classified into two types, namely convecting and non convecting type. The flow chart of solar pond classification is given in Fig.7.25.

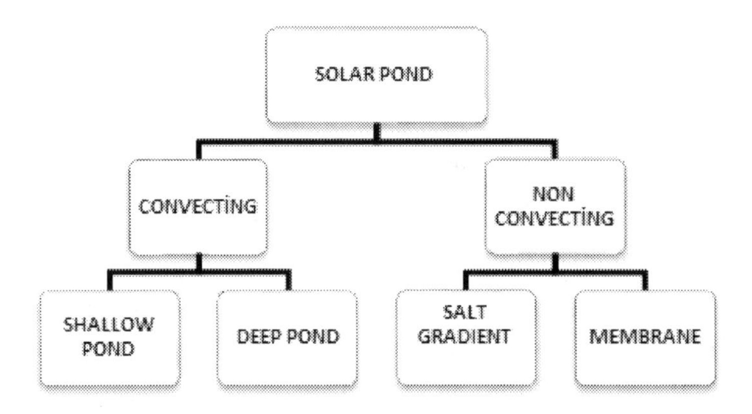

Fig.7.25. Classification of solar pond

a. Convecting solar ponds (Shallow ponds)

A shallow solar pond consists of a group of collectors made of black plastic liners lying on top of insulation that has been laid on flat graded ground. Above the water bag is at least one translucent tarpaulin supported by side curbs. When there is sufficient sunshine, water is pumped into the collectors from an underground reservoir. The water inside the collectors is heated by solar radiation absorbed by the pond liner and can reach temperatures of up to 60°C (140°F). Once the water in the collectors is heated, it can be pumped to an industrial outlet or to a hot storage tank for later use. At night or during periods of low solar radiation, the water in the collectors can be drained back into the underground tank, saving the collected heat.

b. Convecting solar ponds (Deep ponds)

» Salt-free pond.

» Differs from shallow solar ponds only in that water does not need to be pumped to and from the reservoir.

» Double glazing is covered by deep salt-free lakes.

» When solar energy is not available, placing insulation on the glazing reduces heat loss.

c. Non convecting solar pond (Membrane ponds)

» Prevents convection by physically separating the layers with thin transparent membranes.

» As with lakes with a salinity gradient, heat is removed from the bottom layer.

d. Non convecting solar pond (Salt gradient ponds)

When an open body of water absorbs solar energy, convection currents are formed. When the sun's rays that pass through the surface layer are absorbed in the lower layers, this water is heated and rises to the surface, where the heat is transferred by convection to the surrounding air. As the water cools, the density increases and the surface water moves downward. This movement of water balances the temperature throughout the body of water. A salt gradient solar pond uses a salt concentration gradient to suppress natural convection. Heated water can hold more dissolved salt than colder water. Salty, heated water is also heavier and thus remains at the bottom of the solar pond. Solar radiation penetrating the upper layers of the lake is absorbed at the bottom and captured by a non-convective fall layer, which acts as an effective heat insulator against convection.

7.2.5.2. Applications of solar ponds

» Salt production

» Aquaculture, using saline or fresh water

» Dairy industry

» Fruit and vegetable canning industry

» Grain industry

» Water supply

7.2.6. Solar distillation

The sun's heat is used to turn ocean water into vapor and precipitate on Earth in the form of fresh water. Due to rapid population growth, accelerated industrial growth and increased agricultural production, the demand for fresh water continues to increase. The demand for fresh water (drinking water) has increased from 15-20 liters/person/day to 75-100 liters/person/day. The ocean covers 71 recent parts of the earth's surface - 140 million square miles with a volume of 330 million cubic miles and an average salt content of 35,000 ppm. Brackish/saline water is strictly defined as water with less dissolved salts than seawater, but more than 500 ppm. Artificial solar distillation is used to meet the demand for fresh water. The principle of operation of the solar still is shown in Fig. 7.26.

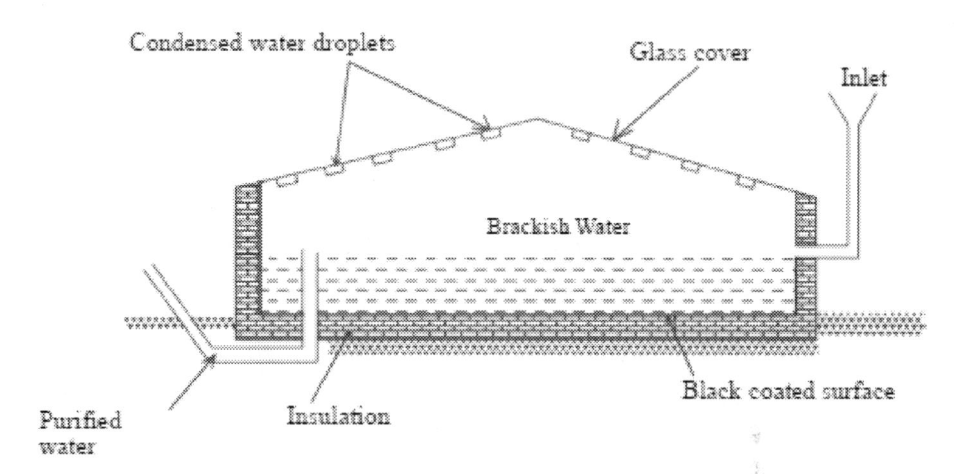

Fig.7.26. Solar still

A solar still is a means of forcing moisture condensation out of the ground and vegetation. Solar works even during the day. As the sun heats the vegetation inside, the leaves sweat the moisture. This condenses on the plastic and then runs down into the cup. The solar still can also be used as a dew pit overnight.

7.2.6.1. Main problems of solar still

» Low distillate performance per unit area

» Vapor leakage through joints

» High maintenance

» Productivity declines over time for various reasons

» The cost per unit of output is very high

Chapter - 8

Wind Energy

Winds are air masses in motion and the primary energy source behind this movement is the sun. The earth receives short wave radiation from the sun and the amount of energy so received is higher at the equator and lower latitudes than at poles and higher latitudes. At the same time the earth losses its heat through long wave radiation more from higher latitudes than from lower latitudes. Consequently, there is a net gain in heat south of 38°N. This is in the northern hemisphere. The atmosphere tries to restore thermal equilibrium by transporting heat from tropics towards the poles. This is the basis for large-scale circulation.

8.1. Wind Controlling Forces

There are three main parameters which affect the wind are, pressure difference, Coriolis force and friction.

8.1.1. Pressure difference

The temperature difference results in pressure difference, which is the main driving force for winds. When the equator region on the earth surface is heated up, the air present above the equator gets heated up and it reduces its density and it tends to move upward. To equate the pressure, cold air from north and south poles will move towards the equator region. This causes the wind to flow.

8.1.2. Coriolis force

Coriolis forces resulting from the rotation of earth, the moving air is deflected to its right side in the northern hemisphere.

8.1.3. Friction

Friction has an important effect on the wind only in the first few kilometers of the Earth's surface. Friction slows the movement of air. By slowing the movement of air, friction also reduces the Coriolis force, which is proportional to wind speed.

8.2. Types of Winds

Three different air circulation systems are primary, secondary and tertiary and are shown in Fig.8.1

1. The primary winds: trade winds, westerlies and polar winds.
2. The secondary winds : monsoon, cyclones and anticyclones
3. Tertiary winds : local winds

8.2.1. Primary winds

These winds constitute large scale motion of atmosphere under the influence of pressure gradients, Coriolis effect, and frictional force. It is related to general circulatory pattern of winds on a rotating earth's surface. It ignores seasonal heating and land water contrast on the earth's surface.

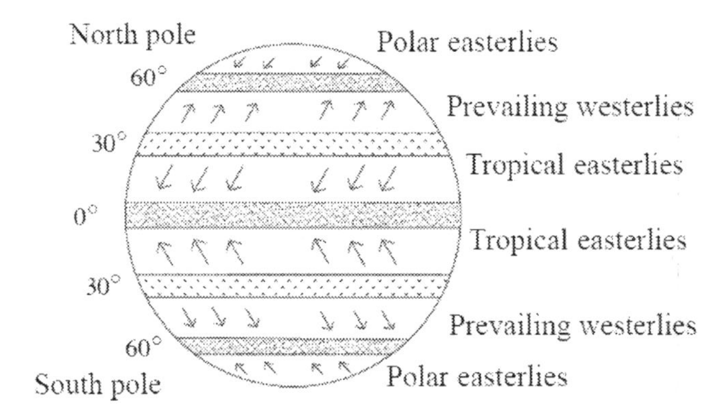

Fig.8.1. Types of wind

8.2.2. Secondary winds

The winds that change their direction periodically with the change in season are called secondary or periodic winds. It depends on monsoon, air masses and fronts, cyclones and anticyclones.

8.2.3. Tertiary winds

Tertiary winds are generated by immediate influence of the surrounding terrain. They are of environmental importance in various ways and exert powerful stress on animals and plants when winds are dry and extremely hot. The land and sea breezes, as well as mountain and valley breezes are also one class of local winds. These winds respond to local pressure gradients set up by heating or cooling of the lower atmosphere.

8.3. Power Law

Wind speed increases with height. The variation of wind speed in the atmospheric boundary layer from a zero value close to the ground to a value approaching the speed of free flowing winds in higher layer can be expressed by a law called 'Power Law'. The Power Law states that

$$V_1 / V_2 = (h_1 / h_2)^a$$

Where, V_1 = Wind speed at height h_1

V_2 = Wind speed at height h_2 and

a = power law coefficient

In well-exposed sites, such as airports, characterized with low surface roughness, a power law exponent of 1/7 (a=1/7) is usually applied. The wind speeds have traditionally been measured at a standard height of 10 m. At this height, wind velocity is found to be 20-25% greater than that close to the surface. At a height of 60 m, they may be 30-60% higher because of the reduction in the drag effect of the earth's surface.

In areas with high ground roughness such as in crowded cities, the value of 'a' is taken as 0.4. For this, the wind speed at 10 m height will be 52% of the wind speed at 50 m height.

8.3.1. Theory of wind power

This theory states that the wind power produced is proportional to the cube of the wind speed. It is also proportional to the square of the diameter of the rotor system (or) wind energy converter. The power in the wind can be computed by using the concept of kinetics. The windmill works on the principle of converting kinetic energy of the wind to mechanical energy. Let ρ be the density of air and m the mass of air passing in unit time through an area A.

Then $m = \rho AV$

Then, kinetic energy or available wind power Pa is,

$$Pa = \frac{1}{2} mv^2$$
$$= \frac{1}{2} (\rho AV) V^2$$
$$= \frac{1}{2} \rho AV^3$$

Therefore, wind power Pa is proportional to

--Density of air
--Intercepted area
--Cube of velocity

Also, if D is the diameter of the horizontal wind energy converter, then

$$A = (\pi/4)D^2$$

Therefore $Pa = \frac{1}{2}\rho((\pi/4) D^2V^3$

$$= (1/8) \rho \pi D^2V^3$$

Therefore the maximum available wind power is proportional to

--the square of the diameter of the rotor.

--the cube of the wind velocity.

8.3.2. Betz coefficient

Betz (1919) showed that the maximum fraction of the kinetic energy that can be extracted from the wind is 16/27 or 0.593. It is also known as Betz Coefficient.

Theoretical maximum power output = $0.593 (1/2 \rho AV^3) = 0.297 \rho AV^3$

The process of energy conversion leads to power reduction which varies with the type of wind turbine and aero generator. It is roughly 1/3 of the theoretical maximum power output.

Available power output $= 2/3 (0.297\rho AV^3)$

$$= 0.198\rho A V^3$$

$$= 0.2 \rho AV^3$$

8.3.3. Power Co-efficient (Cp)

It is the fraction of the free flow wind power that can be extracted by a rotor. In other words, it is the ratio of the power output of wind rotor to the maximum power available in the wind.

$$\text{Power Co-efficient (Cp)} = \frac{\text{Power output of wind rotor}}{\text{Power available in the wind}}$$

Maximum power output of wind rotor = $0.593\,(1/2\rho AV^3)$

Maximum power available in the wind = $\frac{1}{2}\rho AV^3$

$$\text{Maximum value of power coefficient (CPmax)} = \frac{0.593\,(1/2\,\rho AV^3)}{\frac{1}{2}\rho AV^3}$$

$$= 0.593$$

8.3.4. Overall Power Coefficient (COP)

It is the ratio of the power output at generator or pump to the available power in the wind.

$$\text{COP} = \frac{\text{Power output at generator or pump}}{\text{Available power in the wind}}$$

$$= \frac{0.2\,(\rho AV^3)}{\frac{1}{2}(\rho AV^3)}$$

$$= 0.4$$

8.3.5. Characteristics of a suitable wind power site

a. A site should have a high annual wind speed

b. There should be no tall obstruction for a kilometer or two in the upwind side.

c. The top of a smooth well-rounded hill with gentle slopes lying on a flat plain or located on an Island in a lake or sea

d. An open plain or an open shoreline

e. A mountain gap that produces wind tunnelling

8.4. Windmills – Types – Principles

Wind turbines fall into two general types: horizontal axis and vertical axis. A horizontal axis machine has blades rotating on an axis parallel to the ground. A vertical axis machine has blades rotating on an axis perpendicular to the ground. There are a number of designs available for both, and each type has certain advantages and disadvantages. However, compared to the horizontal axis type, very few vertical axis machines are commercially available.

1. Vertical axis wind turbines, where the axis of rotation is vertical to the ground (and roughly perpendicular to the wind current),
2. Horizontal axis turbines, where the axis of rotation is horizontal to the ground (and roughly parallel to the wind flow.)

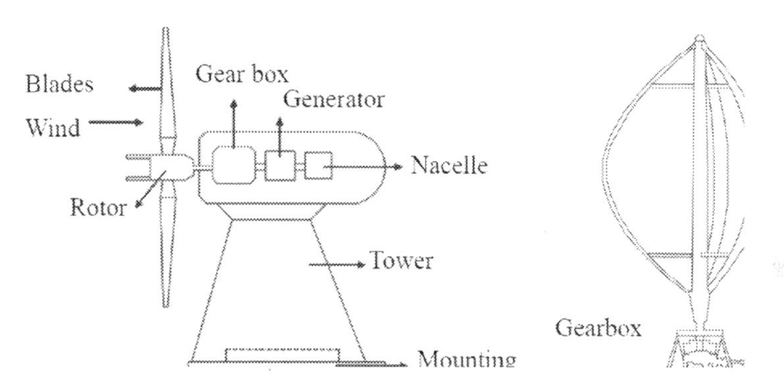

Fig.8.2. Wind turbine configuration

Fig. 8.2. illustrates two types of turbines and typical subsystems for a power generation application. Subsystems include a blade or rotor that converts wind energy into rotating shaft energy; a drive train, usually with a gearbox and generator; a tower that carries the rotor and drive train; and other equipment, including controls, electrical cables, ground support equipment and interconnecting equipment.

8.4.1. Horizontal axis

This is the most common wind turbine design. In addition to being parallel to the ground, the blade's axis of rotation is parallel to the wind flow. Some machines are designed to operate in upwind mode, with vanes upwind of the tower. In this case, a tail vane is usually used to point the vanes into the wind. Other designs operate in downwind mode, so the wind passes the tower before hitting the blades. Without a tail vane, the rotor of the machine naturally follows the wind in downwind mode. Some very large wind turbines use a motor-driven mechanism that turns the machine in response to a wind direction sensor mounted on the tower.

8.4.2. Vertical Axis

Although vertical axis wind turbines have been around for centuries, they are not as common as their horizontal counterparts. The main reason is that they don't take advantage of higher wind speeds at higher altitudes like horizontal axis turbines. The basic vertical-axis designs are the Darrieus, which has curved blades, the Giromill, which has straight blades, and the Savonius, which uses blades to catch the wind. A machine with a vertical axis does not need to be oriented with respect to the direction of the wind. Since the shaft is vertical, the gearbox and generator can be mounted at ground level, allowing for easier servicing and a lighter and cheaper tower. Although vertical axis wind turbines have these advantages, their designs are not as efficient at gathering energy from the wind as horizontal machine designs.

8.5. Wind Energy Conversion system

A wind energy conversion system consists of three main aspects; aerodynamic, mechanical and electrical aspects. The electrical aspect of the WECS can be further divided into three main components, which are wind turbine generators (WTG), power electronic converters (PEC) and the distribution network, as shown in Fig.8.3.

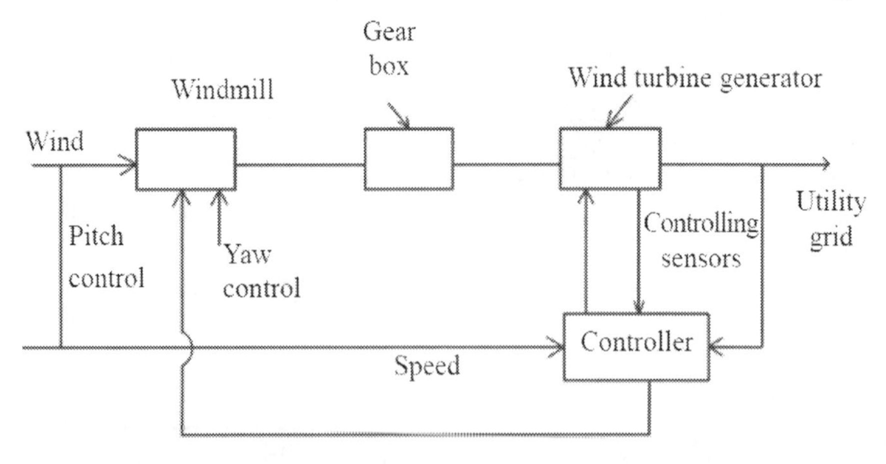

Fig.8.3. Wind Energy Conversion System

8.5.1. Classification of WEC Systems

1. There are two broad classifications:

i) *Machines with a horizontal axis.* The axis of rotation is horizontal and the plane of the aeroturbine is vertical to the wind.

ii) *Machines with a vertical Axis.* The axis of rotation is vertical. The sail or blades may also be vertical, as on the ancient Persian windmills, or nearly so, as on the modern Darrieus rotor machine.

2) Then, they be classified *according* to size as determined by their useful electrical power output.

i) *Small Scale* (upto 2 kW). These might be used on farms, remote applications, and other places requiring relatively low power.

ii) *Medium Size Machines* (2-100 kW). These wind turbines may be used to supply less than 100 kW rated capacity, to several residences or local use.

iii) *Large Scale or Large Size Machines* (100 kW). Large wind turbines are those of 100 kW rated capacity or greater. They are used to generate power for distribution in central power grids. There are two sub classes:

a) Single Generator at a single site.

b) Multiple Generators sited at several places over an area.

3) As per the type of *output power*, wind aero generators are classified as:

i) DC output

a) DC generator

b) Alternator rectifier

ii) AC output

a)Variable frequency, variable or constant voltage AC.

b) Constant frequency, variable or constant voltage AC.

4) As per the *rotational speed* of the aero turbines, these are classified as:

i) *Constant Speedwith variable pitch blades*: This made implies use of a synchronous generator with its constant frequency output.

ii) *Nearly Constant Speed with fixed pitch blades*: This mode implies an induction generator.

iii) *Variable Speed with fixed pitch blades:* This mode could imply, for constant frequency output:

a) Field modulated system

b) AC-DC-AC link

c) Double output induction generator

d) AC Commutator generators

e) Other variable speed constant frequency generating systems.

5) Wind turbines are also classified as per how the *Utilization of output is made:*

i) Battery storage

ii) Direct connection to an electromagnetic energy converter

iii) Other forms (thermal, potential etc.) of storage

iv) Interconnection with conventional electric utility girds.

The system engineer seeking to integrate WECS will, naturally be most interested in the latter case but should be aware of the WECS offer other options as well.

8.5.2. Wind Turbine Generators (WTG)

Wind Turbine Generators can be classified into two major types according to its operation speed and the size of the associated converters as below:

» FSWT (Fixed Speed Wind Turbine)

» VSWT (Variable Speed Wind Turbine) with:

- Narrow speed range
- Wide speed range

The electrical systems of wind turbine generators can be divided into two main groups - fixed speed and variable speed. Fixed speed wind turbines feature a generator connected directly to the grid and are not the most common type today.

They have major advantages over variable speed turbines, having relatively simple build and low price of the electrical system.

Variable-speed wind turbines are the dominating type of turbines. Especially in the high power turbines range, usage of variable-speed turbines gives increased power quality, noise reduction and reduced mechanical stress on the wind turbine. Variable speed wind turbines are equipped with a power converter that allows the frequency of the generator to be different from the grid frequency.

8.5.2.1. Fixed-speed wind turbines

Early prototypes of fixed speed wind turbines used synchronous generators. Due to the advantages of induction machine, such as lower cost, better environmental resistance and excellent mechanical compatibility with rapid wind changes, fixed speed wind turbines are used with induction generator type and are directly connected to the grid. In most wind turbine designs, the generator is connected to the hub with the blades via a gearbox. They are located in a gondola at the top of the turbine tower. A gearbox is needed to change the low rotational speed of the turbine to a high rotational speed on the generator.

Almost all manufacturers of fixed speed wind turbines use induction generators connected directly to the grid. In order to avoid a large inrush current, a soft starter is used for turbine start-up sequence, limiting the current escalation. Since the frequency of the grid is fixed to 50 or 60Hz (depending on the regulations), the speed of the turbine is controlled by the gearbox gears ratio and by the number of poles of the generator. In order to increase the amount of output power, some fixed-speed turbines designs are equipped with a two speed generator and this way they can operate at two different speeds.

8.5.2.2. Variable speed wind turbine

There are many similarities in major components construction of fixed-speed wind turbines and wind turbines operating within a narrow variable-speed range. Fixed-speed wind turbines operating within a narrow speed range usually use a double-fed induction generator and have a converter connected to the rotor circuit.

Nowadays, many wind turbine manufacturers are using variable speed wind turbine systems. The electrical system for variable speed operation is a lot more complicated, in comparison to fixed speed wind turbine system. The variable speed operation of a wind turbine can be obtained in many different ways with differentiation for a broad or a narrow wind speed range. The main difference between wide and narrow wind turbines speed range is the energy production and the capability of noise reduction. A broad speed range gives larger power production and causes reduction of noise in comparison to a narrow speed range system. One of the biggest advantages of variable speed systems controlled in a proper way is the reduction of power fluctuations emanating from the tower shadow.

8.5.2.3. Narrow speed range wind turbines

Several large manufacturers use a doubly fed induction generator with an inverter connected to the rotor circuit in a narrow speed range electrical system. The second option is to use a variable rotor resistance.

8.5.2.4. Wide speed range wind turbines

In broad range variable speed systems, the alternating current from the generator needs to be rectified first and then inverted again into alternating current, before it is going to be fed into the electrical system. Such systems are equipped with a frequency converter. The electrical system of such a wind turbine must consist of three main parts – a synchronous generator, a rectifier and an inverter. The generator and the rectifier have to be chosen as a set, while the inverter can be chosen almost independently of the generator and rectifier parameters. There is a design solution enabling to omit the usage of the gearbox in a broad range variable speed turbine system. It is made by placing a large sized, direct driven multipole synchronous generator. The generator excitation can be made electrically or by permanent magnets.

8.5.3. Power electronic converters

In several system concepts, the generated active power must be adapted in voltage and frequency to the output or consumer side. This is especially the case in variable speed systems. The devices serving for this purpose are power electronic inverters.

Depending on the task the following kinds of inverter are distinguished:

a) AC/DC inverters (rectifiers)

They transform alternating current of a given voltage, frequency and number of phases into direct current. Uncontrolled devices contain diodes, normally in bridge arrangement. Most used are the two-pulse (Graetz) bridge for single-phase input, and the six-pulse bridge for three-phase input. Controlled AC/DC inverters are used in external commutated as well as in self-commutated schemes.

b) DC/AC inverters

They transform direct current into alternating current of a certain voltage, frequency and number of phases. These devices are either external or self-controlled. When the AC side is a grid, these inverters can act also as AC/DC inverters, allowing power exchange in both directions.

c) AC/AC inverters

They transform alternating current of a given voltage, frequency and number of phases into alternating current of another voltage, frequency and number of phases. Inverters with intermediate circuit combine an uncontrolled or controlled rectifier and an external or self-controlled DC/AC inverter. Both inverters are coupled by the intermediate circuit, either with impressed voltage (voltage source- inverter, VSI), or with impressed current (current-source inverter, CSI). Cyclo converters are built without intermediate circuit; the power is transformed in the same inverter A newer concept is the matrix-inverter where fast switches couple input and output side, using no energy storage elements.

d) DC/DC ***inverters (choppers)***

They transform DC current of a given voltage and polarity to DC current of another voltage and polarity. Inverters using an energy storage element and a pulse-control scheme are usually called choppers. Depending on the ratio of output-/input-voltage, we have step-up (boost) inverters for ratios >1 and step-down (buck) inverters for ratios <1. Specific inverter circuits are capable of both ways of operation (buck-boost).

8.6. Energy Storage Devices

Storing electrical energy is an important issue in different technical fields. In power generation and distribution, pump-storage plants are used to supply energy in periods of peak-load, and to pump water back in periods of weak load. Eg., Pumped water, Batteries, Compressed air, Thermal, Flywheel, Superconducting magnetic energy, Hydrogen. Hence storage methods are electrochemical, electrical and mechanical allow reversible energy conversion in both directions, by charging and discharging. Characteristics for applying and comparing different technologies:

» Energy storage capacity

» Power capability

- » Specific values of energy density and
- » Power density

8.6.1. Electrical energy storage devices

8.6.1.1. Double-Layer capacitors

Conventional capacitors are a storage element for electrostatic energy. It is the device for storing energy in larger amounts. It has two electrodes and a fluid electrolyte. When a voltage below that of electrolyte decomposition is applied, ions are collected on the electrodes of opposite polarity in layer thickness of only a few molecules. The very low gaps together with large electrode areas lead to very high values of energy density.

8.6.1.2. Superconducting storage devices

The magnetic field energy is stored by a direct current flowing in a coil which is cryogenically cooled down below its critical temperature by means of the refrigerator. In superconducting state, the coil resistance is zero, so that no current decay occurs and theoretically the magnetic energy is stored indefinitely. The system consists of the superconducting coil, a cryogenically cooled refrigerator, a power electronic rectifier/inverter to convert AC power to the coil arrangement (charging) and to convert DC power back to the AC supply (discharging), and a protection device for emergency discharging.

8.7. Advantages of WECS

i. It is a renewable source of energy

ii. Like all forms of solar energy, wind power systems are non-polluting, so it has no adverse influence on the environment.

iii. Wind energy systems avoid fuel provision and transport.

iv. On a small scale, up to a few kilowatt system is lesser costly. On a large-scale, costs can be competitive with conventional electricity and lower costs could be achieved by mass production.

8.8. Disadvantages of WECS

i. Wind energy availability is dilute and fluctuating in nature.

ii. Unlike water energy, wind energy needs storage capacity because of its irregularity.

iii. Wind energy systems are noisy in operation; a large unit can be heard many kilometres away.

iv. Large areas are needed, typically, propellers 1 to 3 m in diameter, deliver power in the range of 30 to 300 W.

v. Present systems are neither maintenance free nor practically reliable. However, the fact that highly reliable propeller engines are built for aircraft suggests that the present troubles could be overcome by industrial development work.

8.9. Wind Velocity and Direction Measuring Instruments

8.9.1. Dines Pressure - Tube Anemometer

It produces an augmentation or diminution of pressure on different portions of fixed or constrained bodies is the basic principle. It is suitable for use on potential wind-power sites and for temporary as well as permanent installations

8.9.2. Cup anemometer

It works on Abbe's effect (Rotation) i.e. A current of air sets in motion a system susceptible of rotation about an axis. It is used for the measurement of wind speed, and are the most extensively used instrument, developed by T.R. Robinson.

8.9.2.1. Counter type cup anemometer

In this instrument three cup rotor of the standard Meteorological office pattern with 5 inch diameter beaded conical cups, drives a counter indicating the wind flow in miles (or kilometers) tenths and hundredths. The mean wind speed is obtained by dividing the measured wind flow by the time taken.

8.9.2.2. Contact type anemometer

The function of these instruments is to measure instantaneous wind speeds in miles per hour as distinct from wind flow, in miles, indicated by the two types described above.

8.9.2.3. Windmill type anemometer

An alternative to the cup anemometer which may have some advantages in response to changing speed wind and in linearity of DC generator with permanent magnets, so that the voltage generated is proportional to wind speed. The voltage is supplied to an indicating, or recording voltmeter.

8.9.2.4. Head anemomter (Vane anemoeter)

It is useful for the measurement of directed whirl free irrational air streams in free jets, at suction and incoming air opening, in ducts and conduits. There are twp types of vane anemometers available, one has digital recorders which records wind run in 10th, 100th and 1000th of mile or kilometers, the second type is connected to a transducer with an amplifier to amplify the impulse which is fed to a recorder.

8.9.3. Recorders used with cup Anemometers

If in wind measurements, anything more than total wind flow, (in mile) which can be obtained from a counter type cup anemometer, is required some form of recorder is necessary for use with either a contact or generator type cup anemometer. Several types of recorder have been developed for different purposes. The recorders may be classified according to what they recorded i) approximately instantaneous ii) short period or iii) long period wind speeds

8.9.4. Wind direction measurement

An essential part of wind direction measuring equipment is a pivoted wind vane which quickly flows direction changes without-overshooting. The direction may be merely observed from the vanes position in relation to fixed arms pointing to the main points of the compass; it may be indicated on a remotely situated, dial pattern instrument or, again, it may be continuously recorded.

8.9.4.1. Direction indicators

These may be directly operated mechanically from the vane, under which they are located, but more usually, and conveniently, they are operated electrically. The latter type requires a transmitter driven by the vane-which must be capable of producing the necessary torque and a receiver to act as indicator.

8.9.4.2. Direct recorders

Direction recording by the Dines anemometer has already been described.

8.10. Analysis of Aerodynamic Forces Acting on Blades

The Fig.8.4 illustrates the basic aerodynamic operating principles of a horizontal axis wind turbine.

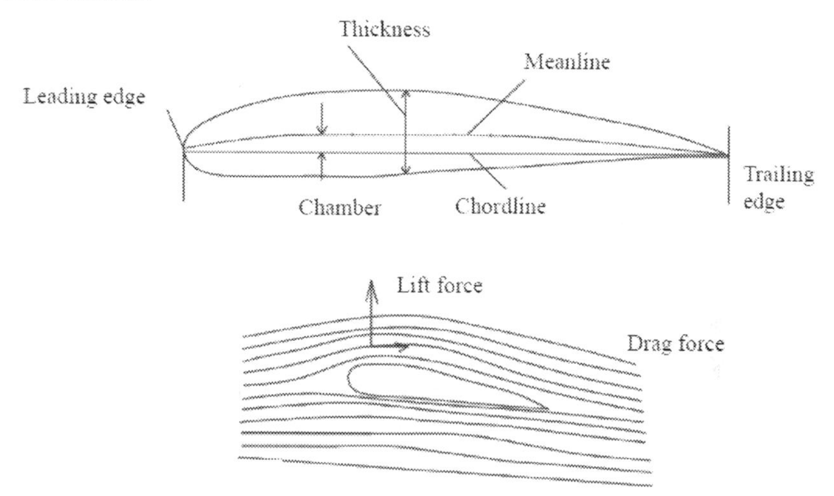

Fig.8.4. Aerodynamic operating principles of air foil

The wind passes over both surfaces of the blade in the shape of a profile. It passes faster over the longer (top) side of the airfoil, creating an area of lower pressure above the airfoil. The pressure difference between the upper and lower surfaces results in a force called aerodynamic lift. In an airplane wing, this force causes the airfoil to "lift" and lift the airplane off the ground. Because the wind turbine blades are forced to move in a plane with the hub as its center, the lift force causes rotation around the hub. In addition to the lift force, a "drag" force perpendicular to the lift force prevents the rotor from rotating. The main goal in wind turbine design is for the blade to have a relatively high lift-to-drag ratio. This ratio can be varied along the length of the blade to optimize the energy output of the turbine at different wind speeds.

8.11. Applications of Wind Energy in Agriculture

8.11.1. Water pumping windmills

Water supplies such as wells and dugouts can often be built in an open paddock. However, the availability of energy sources in the open field is often limited, so some alternative form of energy is required to transport water from the source to the point of consumption. Wind energy is an abundant source of renewable energy that can be exploited for pumping water in remote locations, and windmills are one of the oldest methods of harnessing the energy of the wind to pump water. The typical water lifting mill is represented in Fig.8.5.

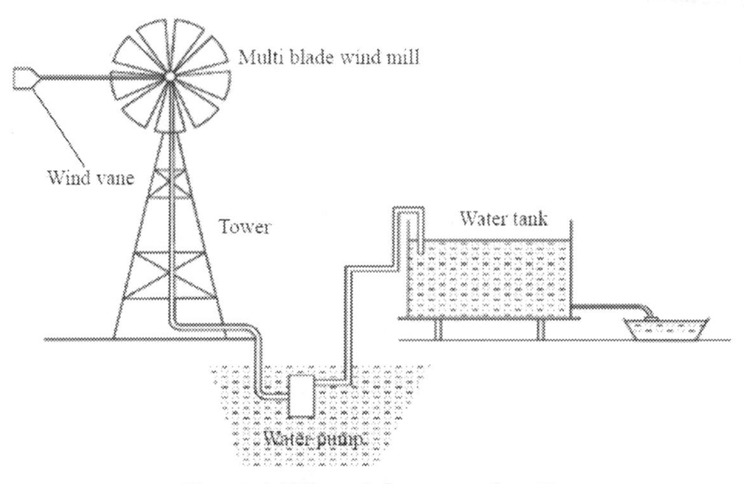

Fig.8.5. Water lifting wind mill

a) Quantity of water deliverable from wind-powered pump

The amount of water that a wind-powered water pumping system can deliver depends on the speed and duration of the wind, the size and efficiency of the rotor, the efficiency of the pump used, and how far the water must be lifted. The power delivered by a windmill can be determined from the following equation:

$$P = 0.0109D^2V^3\eta$$

Where, P is the power in watts, D is the diameter of the rotor in meters, V is the wind speed in kilometers per hour and η is the efficiency of the wind turbine. As can be seen from this expression, relatively large increases in power result from relatively small increases in rotor size and available wind speed; doubling the size of the rotor will result in a fourfold increase in power, doubling the wind speed will result in an eightfold increase in power. However, the efficiency of wind turbines drops significantly in both low and high winds, so most commercially available windmills work best in the wind speed range between about 15 km/h and 50 km/h.

b) Pumps

If the windmill is used to generate electricity to power an electrically driven pump, it is likely that the electricity will need to be stored in batteries due to the variability of the generation. Therefore, a pump driven by an electric motor for use in conjunction with a windmill that produces electricity should have a direct current (DC) motor. For such systems it is important to use good quality deep cycle batteries and incorporate electrical controls such as blocking diodes and charge controllers to protect the batteries.

The most common type of pump used in windmills is a positive displacement cylindrical pump driven by a reciprocating rod connected to a gearbox on the windmill rotor. The performance of these pumps can be increased by adding springs, cams and counterweights that change the lift cycle and balance the weight of the drive rod, reducing the initial torque and allowing the system to perform better in light winds.

An alternative to the traditional cylinder pump is the air pump. An air pump is a type of deep well pump that is sometimes used to remove water from mines. It can also be used to pump sand and water slurry or other "rough" solutions. In its most basic form, this pump has no moving parts other than a windmill-driven air compressor, and the efficiency of the air compressor is a major factor in determining the overall efficiency of the pump. Compressed air is piped down into the well to a foot connected to the discharge pipe. When air is released into the water column in the discharge pipe, a two-phase mixture of air and water is created that is less dense than the surrounding water in the well. This apparent difference in density causes water to rise in the discharge line.

Air pumps can lift water at rates between 20 and 2,000 gallons per minute, up to about 750 feet. Discharge pipes must be placed deep into the water, from 70% of the pipe height above the water level (for lifts up to 20 feet) to 40% for higher lifts. This is the most significant disadvantage of air pumps, since many wells do not have the required depth of standing water. The advantage of this type of pump is that the windmill can be located outside the well and the windmill/air-compressor combination can also be used to aerate dugouts.

c) Other considerations

The availability of wind power is highly variable and likely to be intermittent. In general, expect an average of 6 to 8 hours per day of water pumping at a rate set for 25 km/h (15 mph) winds. Hand pumping can be done on some windmill pumps in emergencies, but wind-driven pumping units should be used in conjunction with storage facilities capable of meeting three to four days' water demand as a back-up source during periods of low wind.

d) Requirement of wind-powered pumping system

One of the attractions of wind pumping systems is their simplicity and robustness. It is wise to check all rubber diaphragms every year and on compressed air models to check all valves. The anchoring system of the windmill tower should also be checked to ensure that the windmill does not topple over in strong winds. As with any pumping system, a wind powered pumping system needs to be checked regularly to ensure that livestock have an adequate supply of water.

Chapter - 9

New Renewable Sources of Energy

9.1. Hydro Power

Hydropower is the largest renewable energy source. This remarkable achievement can be attributed to the fact that of all renewable energy sources, hydropower is the most reliable, efficient and economical. In addition, the concept of hydropower is quite simple and has been used for a long time. The earliest reference to the use of the energy of falling water is found in the work of the Greek poet Antipater from the 4th century BC. The word "hydro" actually comes from the Greek and means "water". A few centuries later, the Romans were the first to use the water wheel. Thanks to the strong influence of the Romans on Europe through conquest, the waterwheel was soon common across the continent, and by the 1800s tens of thousands of waterwheels had been built. Of course, these early water wheels were not used for power generation, but mostly for grinding crops. Water power was first converted to electricity on September 30, 1882 near Appleton, Wisconsin.

Harvesting energy from water is possible due to the gravitational potential energy stored in the water. As water flows from high potential energy (high ground) to lower potential energy (low ground), the potential energy difference thus created can be partially converted into kinetic and in this case electrical energy by a generator. Basically, there are two main designs that use water to generate electricity: the hydroelectric dam and the pumped storage. The water wheel discussed at the beginning of this article is no longer in use today and has been replaced by a much more economical and efficient dam. Both the water wheel and the dam work on the same general principle, but the dam has the advantage of being more reliable due to the reservoir behind it. The principle is simple: the force of the water released from the reservoir through the dam feeder turns the turbine blades. The turbine is connected to a generator that produces electricity. After passing through the turbine, the water re-enters the river on the lower side of the dam. A schematic representation of the dam power plant is in Fig. 9.1.

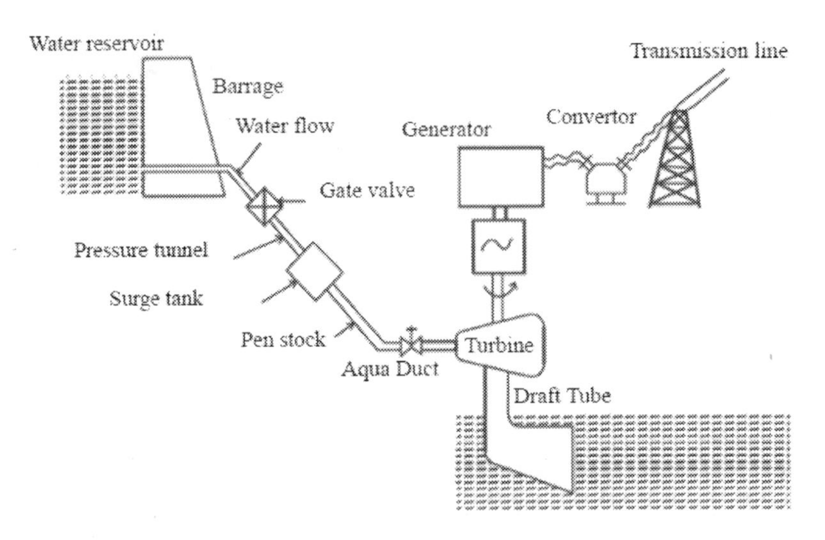

Fig.9.1. Hydroelectric dam based power plant

A pumped-storage plant is very similar to a hydroelectric plant, the main difference being that a pumped-storage plant uses two reservoirs, one of which is significantly higher than the other.

The advantage of this design is that during periods of low electricity demand, such as nights and weekends, energy is stored by reversing the turbines and pumping water from the lower to the upper reservoir. The collected water can later be released to turn turbines and generate electricity as it flows back into the lower reservoir. Now that the two types of facilities have been discussed, there are also two methods of obtaining water: a dam and a runoff river. A dam raises the water level in a stream or river to the elevation needed to create the necessary water pressure. In a run-of-the-river scenario, water is diverted from its natural path, enters a turbine, and is later returned to the river.

Hydropower offers several significant advantages over fossil energy and even other types of alternative energy. Probably the most important currency of hydropower is its reliability. Moreover, it creates no pollution and once the dam is built, although the process is very expensive, the energy produced is practically free. The dam has the ability to generate electricity continuously and can accommodate peaks in demand by storing water above the dam and is able to increase production to full capacity very quickly. Apart from high construction and planning costs, the main disadvantages of large dams are mostly environmental. The dam does not produce harmful emissions as in the case of burning fossil fuels. However, this dramatically changes the landscape and causes several serious, even unbearable changes in the habitat of fish and other plants and animals. Of course, building a large dam will flood a large area of land upstream of the dam, causing problems for the animals that used to live there. It also affects the quantity and quality of water below the dam, which in turn affects plants and animals. Blockage of the river also prevents certain fish migration pattern. Finding suitable sites for dams is also a problem. However, rainfall also affects the ability of dams to generate electricity. Overall, hydropower seems to be a very good source of alternative energy: one that should be maintained as much as possible.

9.2. Geothermal Energy

Geothermal comes from the Greek roots geo, meaning earth, and thermos, meaning heat. Geothermal energy is thermal energy generated and stored on Earth. Thermal energy is the energy that determines the temperature of matter. Earth's geothermal energy comes from the original formation of the planet, from the radioactive decay of minerals (uranium (U^{238}, U^{235}), thorium (Th^{232}), and potassium (K40)), from volcanic activity, and from solar energy absorbed at the surface. The geothermal gradient, which is the temperature difference between the planet's core and its surface, drives the continuous conduction of thermal energy in the form of heat from the core to the surface. The amount of heat within 10,000 meters of the earth's surface is 50,000 times more energy than all the oil and natural gas resources in the world.

9.2.1. Geothermal resources

At any point on the planet, the normal temperature gradient is +300°C per km dug into the ground. So, if one digs down 20,000 feet, the temperature will be about 1900°C higher than the surface temperature. This difference is the normal geothermal gradient. It will be enough to generate electricity. The various geothermal sources are as follows

1. Hydrothermal (Geothermal fields)
2. Brine-methane reservoirs (Geo-pressurized reservoirs)

3. Petro-thermal (Hot dry rock)

4. Exploiting the magma

9.2.1.1. Hydrothermal - Geothermal fields

Geothermal fields are formed when water from rain or snow is able to seep through faults and cracks within the rock, for several kilometers, to reach the hot rock beneath the surface. As the water is heated it rises naturally back towards the surface by a process of convection and may appear there in the form of hot springs, geysers, fumaroles, or hot mud holes as shown in Fig.9.2.

Hot spring (Yellow stone, USA)	Geysers (California, USA)	Fumarol (Hverir, Iceland)

Fig.9.2. Geothermal resources

Most of the geothermal fields that are known today have been identified by the presence of hot springs. Deep geothermal reservoirs may be 2 km or more below the surface, it can produce water with a temperature of 120–350°C. High temperature reservoirs are the best for power generation. Shallower reservoirs may be as little as 100 m below the surface. These are cheaper and easier to access but the water they produce is cooler, often less than 150°C. This can still be used to generate electricity but is more often used for heating. In California, Italy, New Zealand and other countries the presence of these springs led to prospecting usage of boreholes drilled deep into the earth to locate the underground reservoirs of hot water and steam. The Geysers in California are the main fields of this type in use today though others exist in Mexico, Indonesia and Japan. More often the field will produce either a mixture of steam and hot water or hot water alone, often under high pressure.

9.2.1.2. Brine–methane reservoirs (Geo-pressurized Reservoirs)

Geo-pressurized systems are sources of water or brine that has been heated in a manner similar to hydrothermal water, except that this water is trapped in much deep underground aquifers (2400 to 9100 m deep) at relatively low temperature (~ 160°C) and very high pressure (> 1000 bar) with high salinity and is often referred as brine.In some rare cases underwater reservoirs of hot brine are found to be saturated with methane. These reservoirs will normally occur in a region rich in fossil fuel. As reservoir is found it is possible to exploit both the heat in the brine and the dissolved methane gas to generate electricity. The only major reservoirs of this type known today are in the Gulf of Mexico.

9.2.1.3. Petro-thermal (Hot dry rock)

Petrothermal - uses dry hot rocks requiring water injections to make steam. In petro-thermal systems, magma lying close to the earth's heats overlying rock and when no underground water exists, there is simply hot dry rock. The temperatures of hot dry rock vary between 150 to 290 °C. This energy called petro-thermal energy represents by far the largest source of geothermal energy. If water is pumped under high pressure it will cause the rock to fracture, creating faults and cracks through which it can move. If a second borehole is drilled adjacent to the first, then water which has become heated as it has percolated through the rock can be extracted and used to generate electricity.

9.2.1.4. Exploiting the magma

Extracting energy from accessible magma plumes that have formed within Earth's outer crust is the most difficult way of obtaining geothermal energy, but it is also the most exciting because of the enormous quantities of heat available. A single plume can contain between 100,000 MW centuries and 300,000 MW centuries of energy. The easiest geothermal resources to exploit are those that can provide water or steam with a temperature above 200 °C.

9.2.3. Geothermal energy conversion technology

9.2.3.1. Direct -steam geothermal power plant

It occurs where a geothermal reservoir produces high temperature dry steam alone. Under these circumstances it is possible to use a direct-steam power plant which is analogous to the power train of a steam turbine power station but with the boiler replaced by the geothermal steam source. The schematic representation of direct steam geothermal power plant is given in Fig.9.3.

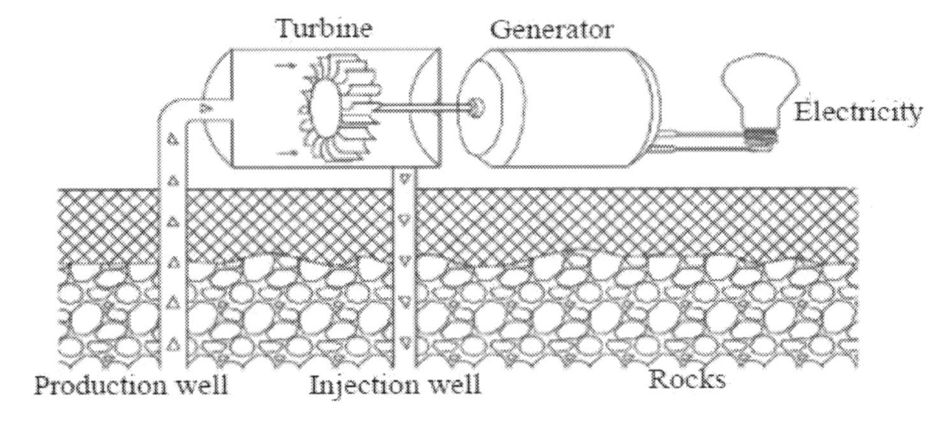

Fig.9.3. Direct steam geothermal power plant

9.2.3.2. Flash-steam geothermal power plant

Most high-temperature geothermal fields produce not dry steam but a mixture of steam and hot water. This is most effectively exploited using a flash-steam geothermal plant. The flash process converts part of the hot, high-pressure liquid to steam and this steam, together with any extracted fluid directly from the borehole, is used to drive a steam turbine. The schematic representation of flash steam geothermal power plant is given in Fig.9.4.

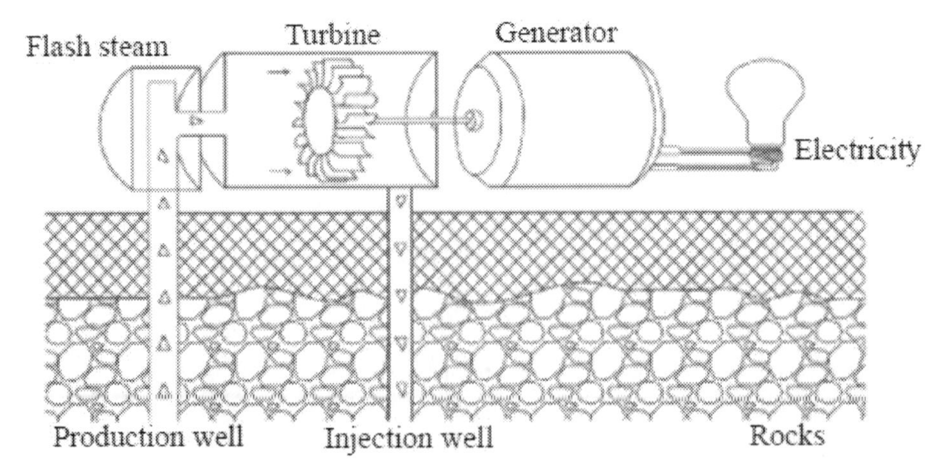

Fig.9.4. Flash steam geothermal power plant

9.2.3.3. Binary geothermal power plant

Where the geothermal resource is of a relatively low temperature a third system called a binary plant is more appropriate. This uses the lower temperature geothermal fluid to vaporise a second low boiling point fluid contained in a separate, closed system. The vapour then drives a turbine which turns a generator to produce electricity. The schematic representation of binary geothermal power plant is given in Fig.9.5.

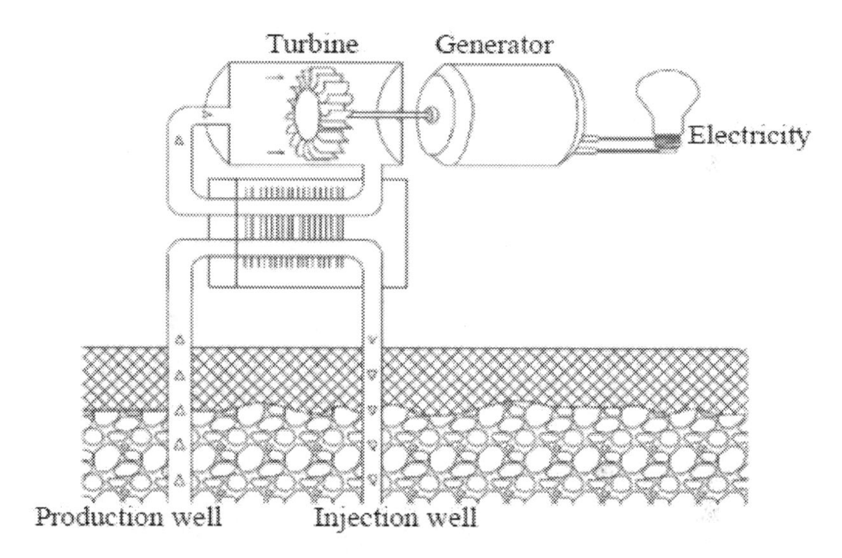

Fig.9.5. Binary geothermal power plant

9.2.3.4. Hybrid geothermal power plant

Some geothermal fields produce boiling water as well as steam, which are also used in power generation. In this system of power generation, the flashed and binary systems are combined to make use of both steam and hot water. Efficiency of hybrid power plants is however less than that of the dry steam plants. The schematic representation of hybrid geothermal power plant is given in Fig.9.6.

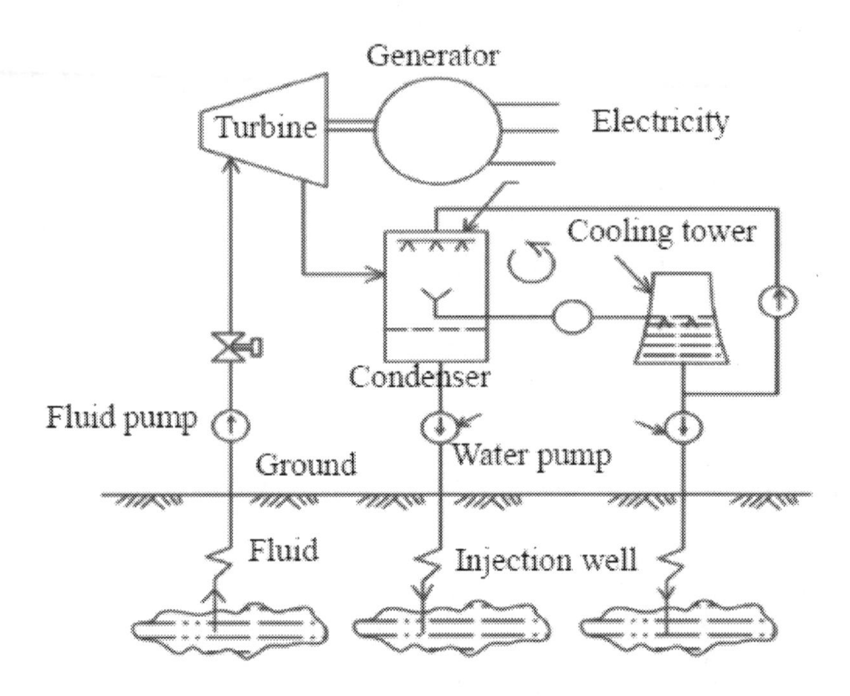

Fig.9.6. Hybrid geothermal power plant

9.2.3.5. Enhanced geothermal system

The term enhanced geothermal systems (EGS), also known as engineered geothermal systems (formerly hot dry rock geothermal), refers to a variety of engineering techniques used to artificially create hydrothermal resources (underground steam and hot water) that can be used to generate electricity. Traditional geothermal plants exploit naturally occurring hydrothermal reservoirs and are limited by the size and location of such natural reservoirs.

9.2.4. Advantages of geothermal energy

Geothermal energy is used to heat homes and generate electricity without producing harmful emissions. The first advantage of using geothermal heat as an energy source is that, unlike most power plants, a geothermal power plant does not produce any pollution and geothermal energy can be used to generate electricity 24 hours a day. Geothermal energy is therefore an excellent source of clean, cheap and renewable energy. If geothermal energy is properly harnessed; does not result in any harmful by-products. Geothermal power plants are generally small and have little impact on the natural landscape or surrounding environment. Because no fuel is used to generate energy from geothermal heat, the operating costs of geothermal power plants are very low. In addition, the cost of land to build a geothermal power plant is usually lower compared to the cost of building an oil, gas, coal or nuclear power plant.

9.2.5. Limitations of geothermal energy

When used incorrectly, geothermal energy can sometimes produce pollutants. Improper drilling into the ground can release dangerous minerals and gases from deep within the earth that can be contained relatively easily. There is also concern that geothermal power plant sites may run out of steam in the long term.

9.3. Wave Energy

Waves are created by the gravitational action of the sun and the moon as also by the force of winds. Oceans are efficient collectors of wind energy; indeed waves are the source of energy steadies than the wind, because once wind generators a wave, the latter conveys the energy it had derived from the wind over immense distance with only moderate dissipation. Wave energy fluxes in the open sea or against wastes may vary from a few watts to kilowatts per meter.

They are smallest in summer and greatest in winter, mainly in the zones of the prevailing waster lies and the trade winds, Wave motion consists of both vertical and horizontal movement of water. Individual particles of water undergo almost a circular motion, moving up as the crest approaches, forward at the crest, down as it recedes, and backward in the trough.

9.3.1. Wave energy generation

The wave energy is generated in two ways. ie. potential energy and kinetic energy. The potential energy arises from the elevation of water above mean level. The equation of wave is given by

$$y=\sin\left(kx - cot\right)$$

Where,

k-$2\pi/\lambda$ = wave number, m^{-1}
w - $2\pi/\lambda$ = Phase rate, $second^{-1}$
y - height above its mean level in m
t - time in seconds
λ - period in seconds

The potential energy per unit area (PE/A) is given as

$$\frac{PE}{A} = \frac{1}{4} gea^2$$

Where,

g = gravitational constant, 9.8 m/s^2

e = water density, kg/m^3

The kinetic energy of the wave is the energy of the water between two vertical planes perpendicular to the direction of wave propagation in x-direction which are one wavelength apart. From hydrodynamic theory, this can be expressed as

$$KE = \frac{1}{4} eg\, pa^2 \lambda L$$

Total Energy and Power density can be written as

$$\frac{E}{A} = \frac{1}{2} gea^2$$

Power = Energy x Frequency

$$\frac{P}{A} = \frac{1}{2} gea^2 f$$

9.4.2. Types of wave energy conversion devices

Wave energy devices can be classified in several ways. However they are mainly classified in two ways: on the basis of location in sea and on the basis of actuating motion used in capturing wave energy. On the basis of location, these devices can be offshore or deep water devices (depth more than 40m) and shoreline devices. On the basis of actuating motion, these devices can be (i) heaving float type, (ii) pitching type and (iii) heaving and pitching float type.

9.4.2.1. Heaving float type

These devices have a float placed on the surface of water which heaves or moves up and down with wave due to rise and fall of water level. The up and down motion of the float is used to operate the piston of an air pump as shown in Fig.9.7. The air pump can be anchored to sea bed. Several floats operated air pumps can be used to store wave energy in the compressed air form which can used to generate electricity with the help of an air turbine coupled to a generator.

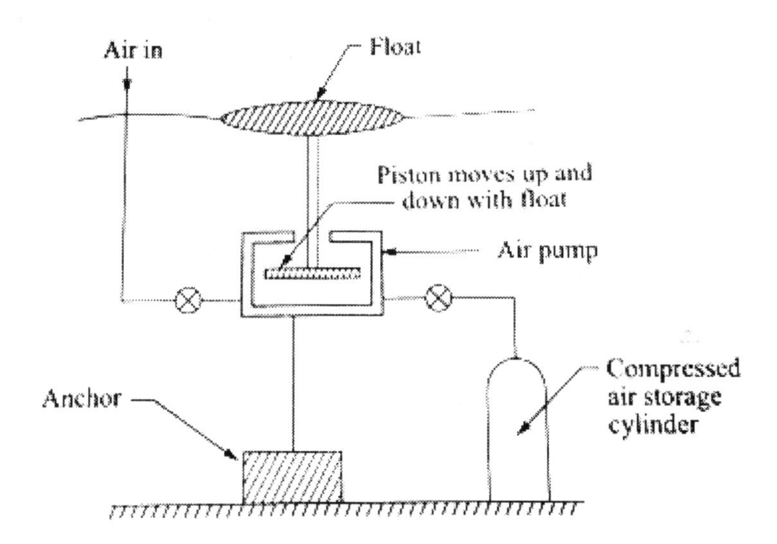

Fig.9.7. Float type wave energy conversion device (air pump)

The motion of float can also be used to run a hydraulic pump so that water can be pumped into an onshore reservoir as shown in Fig. 9.8. The water having higher potential energy can be used to run a turbine. A generator coupled with turbine can generate electric power.

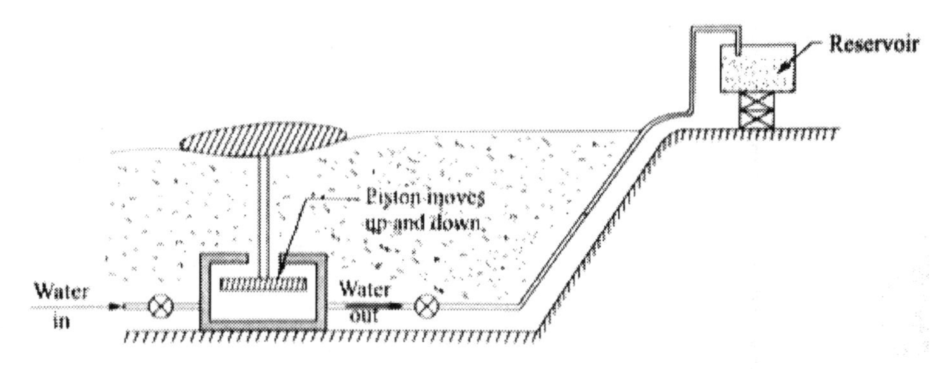

Fig.9.8. Float with hydraulic pump to convert wave energy

9.4.2.2. Pitching type

In pitching type wave energy conversion devices, floating pieces (called nodding ducks) are used to be deflected in the direction of wave motion as shown in Fig. 9.9. These floating pieces or ducks are hinged to a common flexible linkage so that the deflection of these pieces can be converted into rotary motion of a shaft, which is removed by the flexible linkage with a ratchet and wheel mechanism. The floating pieces change direction depending upon the passing of peak or trough of the wave over them. The flexible linkage oscillates with nodding ducks, thereby moving the rotary shaft with ratchet and wheel mechanism. Power collected by the rotary shaft from nodding ducks through flexible linkage helps in driving a generator.

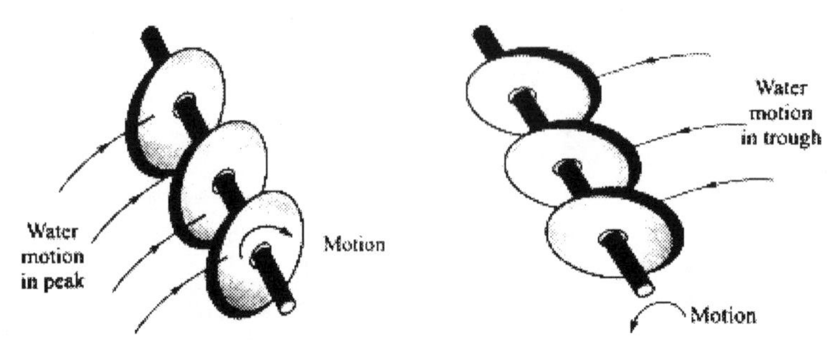

Fig.9.9. Pitching type wave energy conversion

9.4.2.3. Heaving and pitching float type

In these types of wave energy conversion devices heaving and pitching motion of a float is used to extract energy from waves. A specially shaped float known as dolphin rides the wave. The devices use both heaving of connecting rod and pitching motion of the float when waves passes through it. The two motions are converted into unidirectional motion with the help of the rachet and wheel arrangement. The motion is used to operate floating and stationary generators as shown in Fig.9.10.

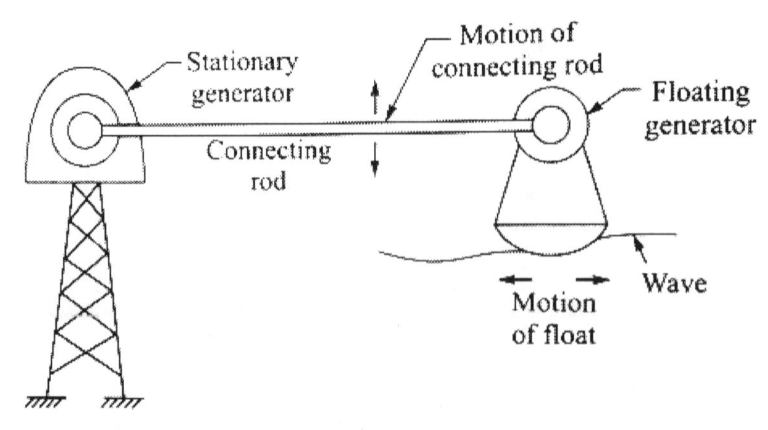

Fig.9.10. Dolphin wave converter using motion of connecting rod and motion of float

9.4.3. Advantages and Disadvantages of wave energy

The advantages are as follows

1. The advantages are as follows:
2. It is a free and renewable energy source.
3. It is pollution free.
4. It is highly suitable to develop power in remote islands, on drilling platforms and on ship where other alternatives are impossible.
5. wave energy conversion devices help in reducing the erosion of costal region.
6. Waves are continuously formed and power can be extracted continuously. No storage of power is required.
7. Wave power devices do not use up large land masses like solar or wind.

The disadvantages are as follows

1. Sea water is corrosive and life of equipment used in conversion devices is limited.
2. Marine growth such as algae adversely affects the working of wave energy conversion devices.
3. Wave energy conversion devices obstruct shipping traffic.
4. Strong waves during storms can damage the wave energy conversion devices.
5. Installation of conversion devices is costly.
6. Repair, replacement and maintenance are difficult to perform.
7. Peak power is available in open sea, where it is difficult to construct, operate and maintain conversion devices as well as transmit power to shore.
8. The slow and irregular motion of wave creates problem to run electrical generator which requires high and constant speed motion.
9. It may cause disturbance to marine life.
10. There is relative scarcity of accessible sites of large wave activity.

9.4. Tidal Energy

Tidal power stations take advantage of the tidal rise and fall to generate electricity. In the simplest type of tidal development, a barrage is built across the estuary of a river. When the tide rises, water flows from the sea into the estuary, passing through sluice gates in the barrage. At high tide the sluice gates are closed and when the tide ebbs, the water behind the barrage is allowed to flow back to the sea through hydraulic turbines, generating power in the process. Exploitation of tidal motion has a long history but its use for power generation is extremely limited with only a handful of operating plants in existence in the world. Tidal electricity generation is only possible in locations where the tidal span (the distance between high and low tide) is significant.

9.4.1. Tidal motion

The motion of the tides is caused primarily by the gravitational pull of the moon and the sun. This motion varies according to a number of cycles. The main cycle is the twice daily rise and fall of the tide as the earth rotates within the gravitational field of the moon. Tidal amplitude in the open ocean is around 1 m. This increases nearer land. Amplitude can be substantially enhanced by the coastal land mass and by the shape of river estuaries The energy that can be extracted from tidal motion waxes and wanes with the tide itself. Power output is generally not continuous. It is, however, extremely predictable. Unlike most other forms of renewable energy, it is not subject to vagaries of weather. This means that the future output of a tidal power station can be determined with great accuracy. The schematic representation of electricity generation from tidal waves is given in Fig.9.11.

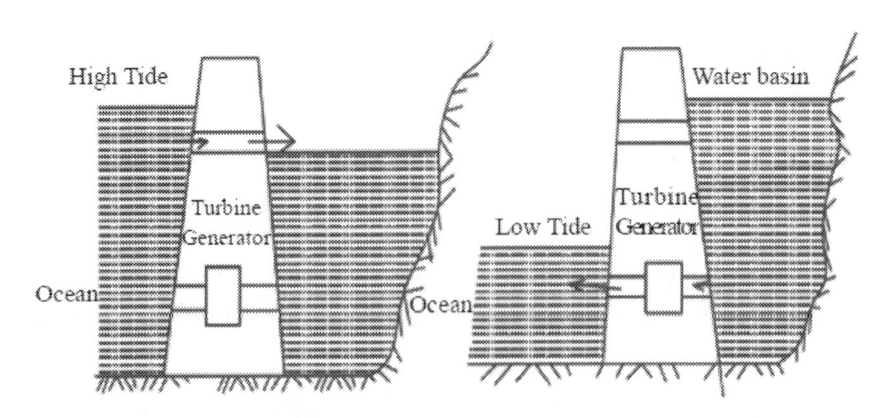

Fig.9.11. Electricity generation from tidal waves

9.4.2. Category of generation

There are several categories of generation for a tidal power plant as given below:

9.4.2.1. Ebb generation

The reservoir is filled through sluices and idle turbines up to high tide. The sluice gates and turbine gates are then closed. They are kept closed until the sea level drops to create enough head over the dike and the turbines generate until the head is low again. Then the sluice gates are opened, the turbines are disconnected and the tank is refilled. The cycle repeats itself. Ebb generation (also known as ebb generation) gets its name because generation occurs as an ebb and flow. Fig. 9.12 represents the ebb generation system.

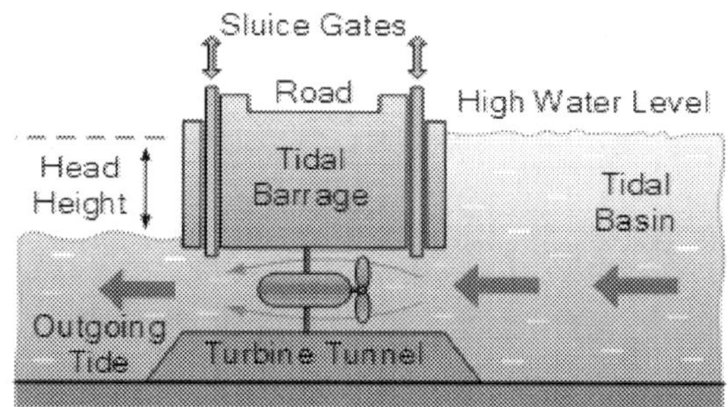

Fig.9.12 Ebb generation system

9.4.2.2. Flood generation

The reservoir is emptied through sluices and turbines created by tidal flooding. This is generally much less efficient than Ebb generation because the volume contained in the upper half of the basin (where Ebb generation operates) is greater than the volume in the lower half (Flood generation domain). Fig. 9.13 represents the flood generation system.

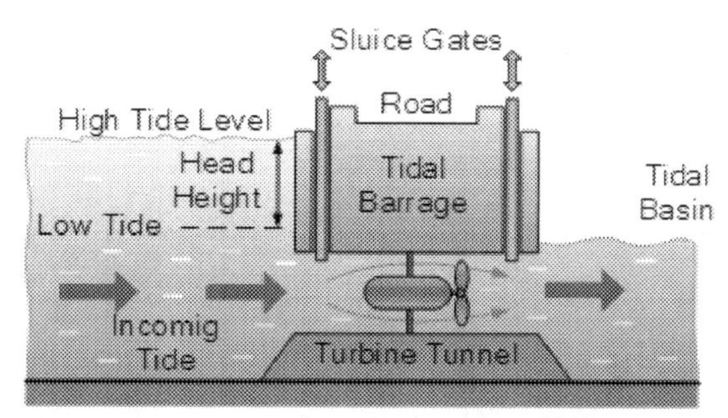

Fig.9.13. Flood generation system

9.4.2.3. Two-way generation

Generation occurs both as tides and floods. This regime is only comparable to tidal generation at spring tides and is generally less efficient. Turbines designed to run in both directions are less efficient.

9.4.3. Advantage of tidal power energy

1. Renewable resource, its maintenance does not need any fuel and is free
2. Totally no pollution, unlike fossil fuels, it produces no greenhouse gases or other waste.
3. Predictable source of energy (compared with wind and solar), it is independent of weather and climate change and follows the predictable relationship of the lunar orbit
4. More efficient than wind because of the density of water
5. It will protect a large stretch of coastline against damage from high storm tides

9.4.4. Disadvantage of tidal power energy

1. Presently costly, very expensive to build and maintain
2. Barrage has environmental affects
   a) Fish and plant migration
   b) Silt and mud deposits
   c) Waste and sewage blocks
3. Technology is not fully developed
4. Only provides power for around 10 hours each day, when the tide is actually moving in or out

9.5. Ocean Thermal Energy Conversion (OTEC)

It relies on the fact that tropical and sub-tropical seas absorb considerable quantities of energy from the sun, energy which elevates the surface water temperature. Deep water does not receive any solar radiation and so remains much cooler. This temperature difference can be used to drive a heat engine. A temperature difference of at least 20°C is normally required to make OTEC effective. This can usually be achieved provided there is a depth of water greater than 1000 m available. OTEC power generation makes use of temperature differences between upper surface layer and deeper layers (800–1000 m) of the sea, generally operating with temperature differences of around 20 degrees centigrade (°C) or more.

This small temperature difference is converted into usable electrical energy through heat exchangers and turbines. First, through a heat exchanger or flash evaporator (in the case of an open-cycle turbine), warm seawater is used to create vapor pressure as the working fluid. The steam then drives a turbine-generator that produces electricity. At the exit of the turbine, the vapor of the working fluid is cooled and condensed back to liquid by cooler ocean water brought in from the deep or seabed. A heat exchanger is also used for this process. The temperature difference before and after the turbine is needed to create a difference in steam pressure in the turbine. The cold seawater used for condensing cooling is pumped from below and can also be used for air-conditioning purposes or for the production of fresh drinking water (via condensation). The auxiliary power required for the pumps is provided by the gross power of the OTEC power generation system.

The advantages of OTEC include the ability to provide electricity continuously (uninterrupted) and at the same time provide cooling without electricity consumption. The capacity factor of OTEC power plants is around 90% - 95%, one of the highest among all power generation technologies. Energy losses caused by pumping are around 20%-30%. The technological problem is that a small temperature difference requires very large volumes of water with minimal pressure losses. This requires large seawater pumps, large piping systems and large cold water pipes operating almost continuously in a hostile and corrosive environment. The cold seawater used in the OTEC process is also rich in nutrients and can be used to cultivate marine organisms as well as plant life near the coast or on land.

9.5.1. OTEC process

Warm water collects on the surface of the tropical ocean and is pumped up using a warm water pump. The water is pumped through a boiler where some of the water is used to heat the working fluid, usually propane or similar material. If it's cooler, you can use a lower boiling point material like ammonia. The propane vapor expands through a turbine that is connected to a generator that produces electrical power. Cold water from the bottom is pumped through the condensers, where the vapor returns to the liquid state. The liquid is pumped back to the boiler. Some small fraction of the energy from the turbine is used to pump water through the system and power other internal operations, but most of it is available as net energy.

9.5.2. Types of OTEC system

9.5.2. 1. Open cycle OTEC or Claude Cycle System

Open-cycle OTEC uses the warm surface water of tropical oceans to generate electricity. When warm seawater is put into a low-pressure vessel, it boils. The expanding steam drives a low-pressure turbine connected to an electrical generator. The vapor that left the salt in the low pressure vessel is almost pure fresh water. By exposure to low temperatures from the deep water, it condenses back to a liquid. In addition, cold seawater pumped from below can be introduced into the air conditioning system after use to facilitate condensation. As such, the systems can produce power, fresh water and air conditioning. In addition, cold water can potentially be used for aquaculture purposes, as seawater from deeper areas near the seabed contains various nutrients such as nitrogen and phosphates. Water is the working fluid and desalinated water can be produced. Fig. 9.14. shows a schematic representation of an open cycle OTEC system.

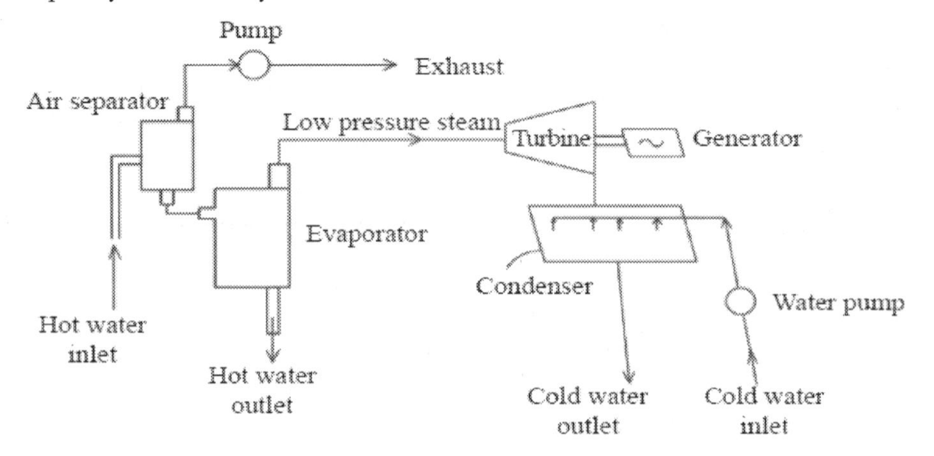

Fig.9.14. Open cycle OTEC system

9.5.2. 2. Closed cycle OTEC or Anderson Cycle System

Closed-cycle systems use a liquid with a low boiling point, such as ammonia, to turn a turbine to generate electricity. Warm surface seawater is pumped through a heat exchanger where the low-boiling liquid evaporates. The expanding steam spins the turbogenerator. Then, cold seawater pumped through a second heat exchanger condenses the vapor back to a liquid, which is then recycled through the system. The main difference between open and closed cycle systems is the much smaller pipe size and smaller turbine diameters for the closed cycle, as well as the area required by the heat exchangers for efficient heat transfer. Closed conversion cycles offer more efficient use of the heat source. Ammonia can be used as a working fluid. Fig. 9.15. shows a schematic representation of the closed cycle of the OTEC system.

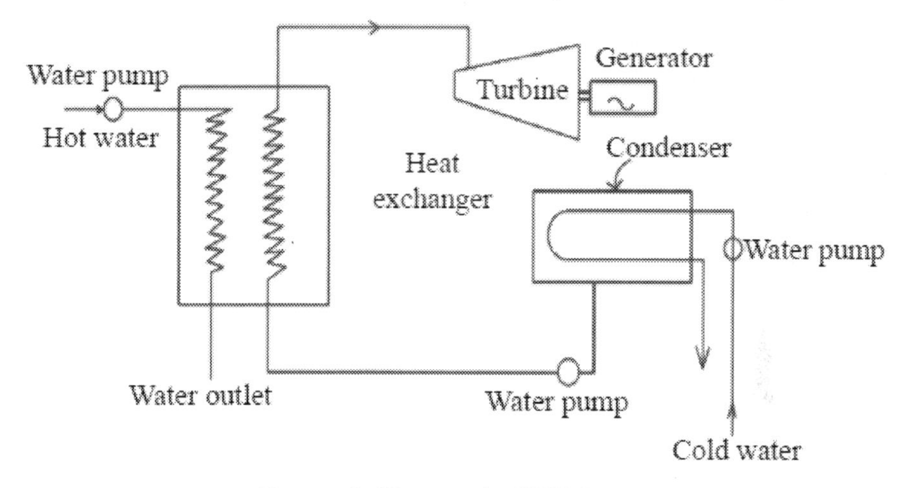

Fig.9.15. Close cycle OTEC system

9.5.2.3. Kalina cycle OTEC system

The Kalina cycle is a variant of the closed OTEC cycle, where a mixture of water and ammonia is used as the working fluid instead of pure ammonia. During evaporation, more heat is drawn into the working fluid and therefore more heat can be converted and efficiency increased.

9.5.2.4. Hybrid system

Hybrid systems combine both open and closed cycle where steam generated by rapid evaporation is used as heat to drive the closed cycle. First, electricity is produced in a closed-loop system as described above. Subsequently, the warm seawater outflows from the closed-cycle OTEC are rapidly evaporated similar to an open-cycle OTEC system and cooled by the cold water outflow. This produces fresh water. Ammonia is the working fluid. Warm seawater is rinsed and then used to evaporate the ammonia.

9.5.3. Different kinds of OTEC Power Plants

There are two different kinds of OTEC power plants,

1. Land based OTEC power plant
2. Floating OTEC power plant

9.5.3.1. Land based OTEC power plant (Onshore plant)

The ground pilot plant will consist of a building. This building will contain heat exchangers, turbines, generators and controls. It will be connected to the ocean by several pipelines and the huge fish farm by other pipelines. Warm water is collected through a shielded cap close to shore. A long pipe laid on a slope collects cold water. Energy and fresh water are produced in the building using the equipment. The used water is first circulated to a marine culture pond (fish farm) and then discharged through a third pipe to the ocean, downstream of the warm water supply. This is done so that the drain does not re-enter the device, as reusing the hot water would reduce the available temperature difference.

9.5.3.2. Floating OTEC power plant

A floating power plant works in the same way as a land-based power plant, the obvious difference being that a floating power plant is floating.

9.5.4. Advantages of OTEC power plants

1. OTEC uses clean, renewable, natural resources. Warm surface sea water and cold deep ocean water replace fossil fuels to generate electricity.
2. Properly designed OTEC plants will produce little or no carbon dioxide or other polluting chemicals.
3. OTEC systems can produce both fresh water and electricity. This is a significant advantage in island regions where fresh water is limited.
4. In the warm layer of the tropical ocean, there is enough solar energy received and stored in the surface layer of the tropical ocean to provide most, if not all, of the current human energy needs.
5. The use of OTEC as a source of electricity will help reduce the state's almost complete dependence on imported fossil fuels.

9.5.5. Disadvantages of OTEC power plants

1. Electricity produced by OTEC would currently cost more than electricity produced from fossil fuels at their current costs.
2. OTEC plants must be located where there is a year-round difference of about 20 °C. Deep oceans must be available relatively close to onshore facilities

for economic operation. Floating plant ships could provide more flexibility.

3. No energy company will put money into this project because it has only been tested on a very small scale.

4. The construction of OTEC power plants and the laying of pipelines in coastal waters can cause local damage to reefs and coastal marine ecosystems.

9.6. Hydrogen as a Source of Renewable Energy

Hydrogen is considered an ideal fuel for future as it has the potential to be used as fuel for running cars and rockets, in meeting the energy needs of buildings and generating the electricity to meet the needs of industry. A hydrogen economy is considered to solve some of the negative effects created while using hydrocarbon fuels owing to the release of carbon during their combustion. Hydrogen is an environmentally friendly source of energy, especially when used in transportation and other applications, because it does not release any pollutant. Free hydrogen does not occur naturally in nature. It has to be generated from some other energy sources. Hence, it cannot be classified as a primary source of energy. Hydrogen is always obtained using some primary source and it is, therefore a secondary source. Hydrogen is a secondary energy carrier just as electricity. As a secondary source of energy, hydrogen is used for the purpose of transportation and storage of primary energy when generated in excess. The primary energy can be a solar or nuclear source which may be harnessed at any remote location. This primary energy can be converted into secondary energy such as hydrogen and the hydrogen can be easily transported to load point where hydrogen can be converted into electricity by fuel cells or into heat by combustion in air.

9.6.1. Properties of hydrogen

Hydrogen has the following properties which make it an attractive alternative energy source

1. It is a light gas. It has a low density which is one-fourth of that of air and one-ninth of that of natural gas.

2. It liquefies at a very low temperature (-253 °C).

3. The specific energy of hydrogen is superior to gasoline on mass basis, but it is inferior on volume basis. The heating value for hydrogen is 120 MJ/kg, while it is 44 MJ/kg for gasoline.

4. It has flame speed during burning which is faster than that of natural gas.
5. It needs lower amount of heat energy to initiate combustion compared to that of natural gas.
6. If can form a combustible mixture with air, with the ratio of 18 to 59%. The mixture can be used for combustion in an IC engine. Hence, the adjustment of air-fuel ratio for IC engine is less critical in case of hydrogen compared to gasoline.
7. It produces water vapour on combustion. Hence, it is pollution free.

9.6.2. Production of hydrogen

The various methods of production of hydrogen include electrolysis of water, thermochemical or steam reforming of methane, thermal decomposition or thermolysis of water and biophotolysis.

9.6.2.1. Electrolysis of water

It is the simplest method of hydrogen production. The method uses an electrolytic which consists of two electrodes immersed in an aqueous conducting solution called electrolyte as shown in Fig.8.1. When a direct current is passed through the cell, it decomposes water into hydrogen and oxygen. Oxygen is formed at anode while hydrogen is formed at cathode. Metal or carbon plates are used as electrodes. The aqueous KOH solution is used as the electrolyte. A decomposition voltage of 2V is applied. The chemical reactions of decomposition of water are as follows:

$$4H_2O + 4e \rightarrow 2H_2\uparrow + 4OH \text{ (cathode)}$$

$$4OH^- \rightarrow O_2\uparrow + 2H_2O + 4e \text{ (anode)}$$

9.6.2.2. Thermochemical method

This method consists of steam reforming of natural gas to produce hydrogen. It is the most efficient, cost-effective and commercialised technology available. The natural gas consisting of methane and carbon monoxide is reformed with the help of steam at 900 °C to produce a mixture of H_2 and CO. CO_2 is removed at the later stage by scrubbing process to get hydrogen. The cost of production of H_2 by this method is very large much same to what it costs to produce electricity using natural gas. The reactions of reforming of natural gas with steam are as follows:

$$CH_4 + 2H_2O = CO_2 + H_2$$

$$CO + H_2O = CO_2 + H_2$$

9.6.2.3. Thermolysis of water

It is process of producing hydrogen by splitting water directly using heat energy. The splitting of water is similar to electrolysis in which electricity is used for splitting and this is the reason why this splitting is called thermolysis. The thermolysis requires a high temperature of about 2500 °C. To carry out thermolysis at the lower temperature of about 850 °C, the process is carried out in different stages using chemical materials which are recovered on completion of the cycle.

9.6.2.4. Biophotolysis

The method uses the ability of plants such algae to generate hydrogen gas when these plants are exposed to water and sunlight. The hydrogen gas can be produced by this method at a low cost. Since this process is essentially a decomposition of water using photons of solar energy in the presence of biological catalysts of plants, the reaction is called biophotolysis of water.

9.6.3. Storage of hydrogen

Hydrogen can be stored in the following forms:

- Hydrogen gaseous form
- Hydrogen liquid form
- Metallic hydrides form
- Chemical hydrides form

9.6.3.1. Hydrogen gaseous form of storage

As hydrogen has lower density compared to natural gas, it requires 3.6 times the volume occupied by natural gas to store the same mass of gas at the same pressure. However, this drawback is compensated as hydrogen can be converted to other forms of energy at the desired location more efficiently compared to natural gas. The hydrogen can be stored in strong steel tanks or cylinders like natural gas, but storage pressure is kept high in the range of 350-750 atm so that sufficient mass of gas can be stored. This method is suitable where a small amount of hydrogen is required. Large amount of hydrogen is generally stored in the underground facilities.

9.6.3.2. Hydrogen liquid form of storage

Hydrogen can be stored in the liquid form as liquid takes less storage space. However, storage space in liquid form requires cryogenic or temperature lower than 20K (-253°C). This low temperature has to be maintained for the complete storage duration, which makes this type of storage a costly preposition.

9.6.3.3. Metallic hydrides form of storage

Hydrogen chemically joins with certain metals and forms hydrides. When hydrides are heated, the hydrogen gas can be liberated. The pressure of gas released depends on temperature and it remains constant at a certain temperature. The metallic alloys used for forming hydrides are lanthanum nickel, iron titanium and magnesium nickel.

The reaction forming metallic hydrides with hydrogen are as follow:

$La\ Ni_5 + 3H_2 \qquad (La\ Ni_5)\ H_6 + heat$

$Fe\ Ti + H_2 \qquad (Fe\ Ti)\ H_2 + heat$

$Mg_2\ Ne + 2H_2 \qquad (Mg_2Ni)\ H_4 + heat$

Owing to the heavy weight of all metallic alloys, hydrides storage is unsuitable to be used for running of vehicles. Metallic hydrides storage is desirable when metal alloys forming hydrides is

(i) inexpensive,

(ii) hydrides is able to contain a large amount of hydrogen,

(iii) storage of hydrogen in metallic alloy can be done at the room temperature

(iv) Significant pressure of liberated gas is possible at the room temperature.

9.6.3.4. Chemical hydride form of storage

In chemical hydride method, chemical compounds are used to form hydrides in alkaline solution. Chemical hydride process is a safer method compared to metallic hydride from of storage. It is also a more volumetrically efficient storage method.

9.6.4. Hydrogen as an ideal source of renewable energy

The Features which make hydrogen an ideal source of renewable energy are as follow:

1. Primary energy resources can be easily stored after conversion as hydrogen.
2. It can be economically and technically stored so that it can be used when required.
3. It can be easily converted into heat on combustion or electricity by using fuel cell.
4. It can be easily transported to long distances through piping system or some other transportation means. Transportation cost of hydrogen is lesser than the cost of transmission of electricity.
5. Its use is pollution free.

Chapter - 10

Energy Plantation

Growing plants intentionally for fuel value is called energy plantations. Energy plantations are designed to provide a significant amount of usable fuel continuously throughout the year at a cost comparable to other fuels. Annual plants, agricultural crops are not suitable for energy plantations as they cannot provide year-round fuel supply and can be better used for other specific purposes such as forest products industry, food industry, fodder etc. Therefore, fast growing trees are best suited. for an energy plantation that grows faster and the energy content will also be higher.

An energy plantation is an economical way of using solar energy. In photosynthetic concepts, the efficiency of converting solar energy into plant material is about 1%, and the overall efficiency of converting solar energy into electricity would be about 0.3%. This is a less efficient conversion method compared to 10% for photovoltaic cells that convert solar energy into electricity. But the advantage of the process of converting solar energy into a plant is that it is a natural process that eliminates sophisticated industrial and technological aspects.

The amount of wood that can be permanently used as fuel in a given area is limited by the local amount of solar radiation. Solar energy is distributed throughout the world, but is concentrated in tropical regions. Tropical regions therefore have the potential to produce more wood per unit area than other regions.

10.1. Selection of Species

The choice of energy plantation species depends on the climate and soils where the energy plantation is designed. The selection of land is based on its suitability for agricultural purposes. If the land is suitable for primary crops and is fertile, it must not be selected for energy plantations. Depending on the soil type and climate of the area, the species and their life cycles, intensive forestry practices allow more trees to be planted per unit area to facilitate smaller spacing between them. The geometry of the leaf-bearing branch should be such that the maximum area faces the sun.

10.2. Importance of Energy Plantation

In addition to the efficient use of unused and unused land, energy plantation has some specific advantages. They are:

10.2.1. Storage

Storage of used solar energy is difficult if it is directly converted into thermal or electrical energy using heat consumers and photovoltaic systems. But the energy plantation is a good system in which the used solar energy can be stored for a longer period of time without any loss; they are also energy accumulators. Energy conversion from plant materials can be carried out over a wider range of periods based on the energy requirement, which is due to the longer maturity period of these types of species.

10.2.2. Economics

Land recovery can be increased. Multiple crops and denser plantations can further increase recovery.

10.2.3. Effects on the environment

Plant matter contains less than 0.1% sulfur, and the problem of SO¬2¬ pollution by burning biomass practically does not exist. The energy conversion can be done thermochemically or biochemically. Whichever one is used, carbon dioxide will be produced. However, this does not disturb the balance, as plants consume carbon dioxide during the growing season. When barren land is transformed into a vegetative cover green belt, clean and health ecological conditions can be resorted.

10.3. Species for Energy Plantations

The choice of plant species depends on the type and quality of the soil, the type and quality of the water, the characteristics of the soil in the given location. Suitable species for firewood / firewood / energy plantations for different regions are as follows

10.3.1. Tropical dry region

Acacia catechu, Acacia modesta, Acacia nilotica, Acacia Senegal, Acacia tortilis, Anogeissus pendula, Albizia lebbek, Azadirachta indica, Cassia siamea, Cordia rothii, Dalbergia sissoo, Emblica officinalis, Eucalyptus camaldulensis, Erythrina superb, Gmelina arborea, Parkinsonia aculeate, Peltophorum ferrugineum, Pongamia pinnata, Prosopis cineraria, Prosopis juliflora, Tamarindus indica, Tamarix troupe, Tecomella undulate and *Zizyphus maurtiana.*

10.3.2. Tropical humid region

Adina cordifolia, Acacia auriculiformis, Acacia catechu, Acacia nilotica, Albizia procera, Azadirachta indica, Cassia siamea, Casuarina equisetifolia, Dalbergia sissoo, Dendrocalamus strictus, Ficus spp., Eucalyptus spp., Kydia calycina, Leucaena leucocephala, Madhuca indica, Melia azedarach, Morus alba, Salix tetrasperma, Syzygium cuminii, Tamarindus indica, Trewia nudiflora, Gliricidia sepium and *Gmelina arborea.*

10.3.3. Sub-tropical region

Acacia catechu, Acacia melanoxylon, Acacia nilotica, Aesculus indica, Ailanthus excels, Celtis australis, Grevillea robusta, Michelia champaca, Populus deltoids, Populus nigra, Robinia pseudoacacia, Salix alba and *Toona ciliate.*

10.3.4. Temperate climate

Acer spp., Aesculus indica, Alnus nepalensis, Alnus nitida, Celtis australis, Populus ciliate, Quercus semecarpifolia, Salix alba and *Toona serrata*

10.3.5. Description of selected species

The description of some of the selected species are as follows

a. Casuarina species

It is a large, evergreen, hardy tree well adapted to coastal areas. These species grow to a height of 15 to 25 m with a trunk diameter of approx. 15-20 cm. In some cases, the height can reach up to 50 m and the diameter up to 30 cm. It grows well in warm subtropical climates. The maturity period is approximately 5-7 years. The yield of chopped form in the second year will be 3-5 tons/ha/year (7-10 m^3/ha/year). Casuarina gave up to 250 tons/ha in four years. The calorific value of the species is about 4900 kcal/kg. This species is considered the best species for firewood. They are used as wood, to make pulp, as protection against wind and against erosion.

b. Eucalyptus species

It is one of the major fast-growing species in India. It grows faster in the early stages i.e. about 15 m in 2-3 years and 30 m in 10 years and reaches a height of about 65 m and a diameter of 2.3 m with a straight trunk and carries 65% of the total weight of the tree. It can therefore also be used on railway or road lines and in the production of wood and oil extraction from leaves. The yield is 60 tons/ha/year or 7-10 m3/ha/year. Calorific value is about 4800 kcal/kg. Some energy species and their yield are listed below in Table 10.1.

Table 10.1. Energy crops spices

Sl.No.	Name of the species	Yield (m^3/ha/yr)	Calorific value (kcal/kg)
1.	Leucaena Leucocephala	30 – 40	4200 - 4800
2.	Prosopis Juliflora	5 – 6	> 5000
3.	Sesbania	15 – 25	
4.	Acacia auriculiformis	10 – 12	4800 – 4900

c. *Acacia nilotica*

Acacia nilotica is an evergreen, medium-sized (height ranging from 2.5 m to 25 m) tree. In favorable locations, it reaches a height of 15 - 25 m and a girth of 2.4 - 3.0 m. In unfavorable locations, it is a dwarf, shrub-like or invasive tree. It has a short, thick and cylindrical trunk. The tree has a very clear trunk up to 5 - 6.5 m high. In difficult habitats, it rarely grows more than 10 m. The clear trunk in such cases is no more than 3 - 4 m.

It includes nine subspecies that differ in pod shape and pubescence and tree habit:indica, kraussiana, leiocarpa, nilotica, subalata and *tomentosa* in Africa, *cupressiformis* , and *hemispherica* in the Indian continent, and a*dstringens* in both continents.

It is widely used in India for firewood and charcoal, also in locomotives and steamships and also in industrial presses. The calorific value of sapwood is about 4800 and heartwood about 5000 kcal/kg. The species also fixes nitrogen.

d. *Glyricidia sepium*

Gliricidia sepium is native to Central America and Mexico. Historically, it has been cultivated as a border fence that does not require any special agronomic practices, care for pest control measures due to some inherited genetic characteristics typical of it. It is a drought-tolerant tree that has the ability to absorb nitrogen from the atmosphere with the least intake of soil nutrients and at the same time bind nitrogen to the soil.

It could be grown as a monoculture or as a mixed crop along with other short/ long term crops grown in the locality. Planting methods include 1m x 1m double row spacing at a density of 8000 trees/ha. After the first harvest, both wood and leaf yields increase, reaching a maximum in the fourth year of planting and then maintaining a wood yield of around 6.0-8.0 kg wood/tree/year (at 20% moisture).

e. *Dalbergia sissoo*

Dalbergia sissoo, commonly known as North Indian rosewood, is an evergreen rosewood, also known as sisu, tahli, Tali and also Irugudujava. It is a medium-sized to large deciduous tree, growing under favorable conditions to a height of 30 m and a diameter of 80 cm. The calorific value of sapwood is 4900 kcal/kg and heartwood 5100 kcal/kg. It is grown alternately for 10 to 15 years as fuel wood. The tree has excellent combing ability, although loss of vitality after two or three rotations has been reported. Wood makes excellent charcoal for heating and cooking.

f. *Albizia lebbeck*

Albizia lebbeck is a species of Albizia, native to Indomalaya, New Guinea and northern Australia. It is a tree growing 18–30 m tall with a trunk 50 cm to 1 m in diameter. Its uses include environmental management, fodder, medicine and timber. It is grown as a shade tree in North and South America. In India and Pakistan, the tree is used for timber production. Albizia lebbeck wood has a density of 0.55-0.66 g/cm^3 or higher. It is an excellent firewood with a calorific value of 5200 kcal/kg.

g. *Pongamia pinnata*

A fast-growing deciduous tree reaching a height of 20 m, believed to be native to India and found throughout Asia. It is a deciduous tree growing to a height of approx. 15-25 m with a massive, equally wide canopy. The leaves are a soft, glossy burgundy in early summer and mature to a glossy, deep green as the season progresses. The tree adapts well to intense heat and sunlight, and its dense network of lateral roots and strong, long taproot make it drought tolerant. Biodiesel can be produced from oil extracted from pongamia seeds and the seed cake can be used as feedstock for biogas plants. The calorific value of wood is 4,600 kcal/kg.

References

Text Books

Kamaraj,S., Angeeswaran, R and Divyabharathi. R. 2014. Text book on Thermochemical Conversion Technologies. A.E Publications. Coimbatore.

Kamaraj,S., Mahendiran, R and G. Saravanapriya. 2013. Solar energy and its applications. AEC&RI, TNAU, Coimbatore, Tamil Nadu, India.

Lehmann, J. 2009. Biochar for environmental management: An introduction. *In: Biochar for environmental management: Science and Technology*. (Eds. Lehmann, J and S. Joseph).Earthscan Ltd, London.pp. 110.

Manwell, J.F.,J.G.McGswan and A.L.Rogers. 2004. Wind Energy Explained – Theory Design and Application, John Wiley and Sons.

Nicholas P. Cheremisinoff.2003. Fundamentals of wind energy. Ann Arbor Science, New York.

Pugalendhi, S and P.Subramanian. 2001. Biogas technology challenges and opportunities. Sri Sakthi Publications, Coimbatore, Tamil Nadu, India.

Pugalendhi, S., N.O.Gopal and P.Subramaninan. 2004. Biogas Technology. TamilNadu Agricultural University, Coimbatore. Pages 224

Rai, G.D. 2002. Non conventional Energy Sources. Khanna Publishers, New Delhi, India.

Rao, S. and B.B. Parulekar. 2002. Energy Technology - Non conventional, renewable and conventional. Khanna publishers, New Delhi, India.

Tany Burtar. 2001. Hand book of wind energy, John Wiley and Sons, New York.

Research Articles

ASTM, 1977. Annual book of ASTM standards part 26, Gaseous fuels; coal and coke : atmospheric analysis. *American Society for Testing Materials, U.S.A.*, 293 – 298 and 369 – 377.

Demiral, I. and E.A. Ayan. 2011. Pyrolysis of grape bagasse : effect of pyrolysis conditions on the product yields and characterization of the liquid product. *Bioresource Technology.*, 102 (4): 3946 – 3951.

Josea, P., Granada, E. Saavedra, A. Eguía,P. and J.Collazo. 2010. Biomass Thermogravimetric Analysis: Uncertainty Determination Methodology and Sampling Maps Generation. *International Journal of Molecular Science. 11* (7): 2701-2714.

Parikh, J., Channiwala, S.A and G.K. Ghosal. 2007. A correlation for calculating elemental composition from proximate analysis of biomass materials. *Fuel.* 86:1710-1719.

Ramesh, D. and A. Samapathrajan and P. Venkatachalam. 2005. Pilot biodiesel plant for vegetable oils. *Periyar Journal of Research and Development,* 3(1): 15-1.

Ramesh, D. and A. Samapathrajan and S. Joshua Davidson. 2005. Fuel Properties of palm oil and its biodiesel production. *Periyar Journal of Research and Development,* 2(3): 25-29.

e-References

www.beeindia.gov.in

www.biochar-international.org

www.iea.gov.in

www.mnre.gov.in

www.teda.in

Printed in the United States
by Baker & Taylor Publisher Services